Studies in Computational Intelligence 455

Editor-in-Chief

Prof. Janusz Kacprzyk
Systems Research Institute
Polish Academy of Sciences
ul. Newelska 6
01-447 Warsaw
Poland
E-mail: kacprzyk@ibspan.waw.pl

For further volumes:
http://www.springer.com/series/7092

Amitava Chatterjee, Anjan Rakshit,
and N. Nirmal Singh

Vision Based Autonomous Robot Navigation

Algorithms and Implementations

 Springer

Authors
Dr. Amitava Chatterjee
Electrical Engineering Department
Jadavpur University
West Bengal
Kolkata
India

Prof. Dr. Anjan Rakshit
Electrical Engineering Department
Jadavpur University
West Bengal
Kolkata
India

Dr. N. Nirmal Singh
Electronics and Communication
Engineering Department
V V College of Engineering
Tuticorin District
TamilNadu
Tisaiyanvilai
India

Additional material to this book can be downloaded from http://extras.springer.com

ISSN 1860-949X
ISBN 978-3-642-42670-4
DOI 10.1007/978-3-642-33965-3
Springer Heidelberg New York Dordrecht London

e-ISSN 1860-9503
ISBN 978-3-642-33965-3 (eBook)

Printed on acid-free paper

Springer is part of Springer Science+Business Media (www.springer.com)

Preface

"Vision Based Autonomous Robot Navigation: Algorithms and Implementations" is devoted to the theory and development of autonomous navigation of mobile robots using computer vision based sensing mechanism. The conventional robot navigation systems, utilizing traditional sensors like ultrasonic, IR, GPS, laser sensors etc., suffer several drawbacks related to either the physical limitations of the sensor or incur high cost. Vision sensing has emerged as a popular alternative where cameras can be used to reduce the overall cost, maintaining high degree of intelligence, flexibility and robustness.

The introductory chapter details the basic concepts of autonomous navigation of mobile robots and the utility of using vision as the sensing mechanism in this context is highlighted. Here a broad categorization of research activities pertaining to vision-based navigation in indoor and outdoor environments is presented. This is followed by an introduction of different broad modalities of obstacle detection and avoidance. In the next chapter, the book discusses how real-life interfacing of external peripherals with a readymade mobile robot can be successfully achieved. Here a detail description of interfacing of such peripherals with the KOALA robot using serial communication in interrupt driven mode is provided. In the next chapter, a vision based robot navigation strategy is detailed, where a subgoal based scheme is employed to follow the shortest path to reach the final goal and also simultaneously achieve the desired obstacle avoidance. This strategy employs a two layer architecture where vision sensor operates in layer 1 and IR sensor based obstacle avoidance scheme operates in layer 2.

The next chapter discusses how a low-cost robot can be indigenously developed in the laboratory with special functionalities. Special emphasis is put on development of two microcontroller based sensor systems for the robot in this regard: (i) an IR range finder system that can be developed with dynamic range enhancement capability and (ii) an optical proximity detector system developed utilizing the principle of switching mode synchronous detection technique. This is followed by the next chapter which presents, in a step-by-step manner, gradually progressing from easier modules to more complex modules, how vision-based navigation subroutines can be actually implemented in real-life, under 32-bit Windows environment.

The next two chapters deal with incorporation of fuzzy logic in the context of mobile robot navigation. Among these, the first one discusses how a vision based navigation scheme can be developed for indoor path/line tracking. Here fuzzy vision-based navigation is hybridized with a fuzzy IR-based obstacle avoidance mechanism. The next chapter first introduces the concept of EKF-based SLAM for mobile robots. Then it discusses a more complex scenario where fuzzy or neuro-fuzzy supervision can be effectively utilized to improve performance for EKF based SLAM in presence of incorrect or uncertain knowledge of sensor statistics. The last chapter discusses how a two camera based vision system can be implemented in reality for SLAM in an indoor environment.

Kolkata, West Bengal, India   Amitava Chatterjee
September 2012   Anjan Rakshit
                                                                   N. Nirmal Singh

Contents

Chapter 1
Mobile Robot Navigation

Abstract. This chapter introduces the basic concepts of autonomous navigation of mobile robots and the utility of using vision as the sensing mechanism in achieving the desired objectives. The chapter discusses the broad categories of vision-based navigation in indoor and outdoor environments. Different prominent directions of research in this context are introduced and also different broad modalities of obstacle detection and avoidance are presented.

1.1 Autonomous Mobile Robot Navigation

Advances in recent technologies in the area of robotics have made enormous contributions in many industrial and social domains in recent times. Nowadays numerous applications of robotic systems can be found in factory automation, surveillance systems, quality control systems, AGVs (autonomous guided vehicles), disaster fighting, medical assistance etc. More and more robotic applications are now aimed at improving our day-to-day lives, and robots are now caught in sight more often than ever before performing various tasks in disguise [1]. For many such applications, autonomous mobility of robots is a mandatory key issue [100]. Autonomous mobile robots are robots which can perform desired tasks in structured or unstructured environments without continuous human guidance. A fully autonomous mobile robot has the ability to:

- Gain information about the environment.
- Work for an extended period without human intervention.
- Move either all or part of itself throughout its operating environment without human assistance.
- Avoid situations that are harmful to people, property, or itself, unless those are part of its design specifications.

An autonomous mobile robot may also learn or gain new capabilities like adjusting strategies for accomplishing its task(s) or adapting to changing surroundings.

1.2 Why Vision in Navigation?

Vision is the sense that enables us, humans, to extract information about the physical world, and, appropriately, it is the sense that humans rely on the most. In recent past, computer vision techniques capable of extracting such information are

A. Chatterjee et al.: Vision Based Autonomous Robot Navigation, SCI 455, pp. 1–20.
springerlink.com   © Springer-Verlag Berlin Heidelberg 2013

continuously being developed and refined. Vision processing is computationally intensive, but as faster and lower priced processors being developed, the development of real-time vision-based navigation systems for mobile robots is becoming a reality for a variety of complicated jobs and more research works are being focused in this domain now, than ever before [100].

The other sensors that are used for navigation include infrared sensors, sonar sensors, laser range finders, the position sensing device (PSD) sensors and inertial sensors. Infrared sensors have limited usage; they are very often used as proximity detectors and the main shortcoming of using them as range finders lies in their limited range and their susceptibility to ambient light interference. IR sensors are also known for their non-linear behavior and their reflectance dependency on the surface of a target [2]. Sonar sensors are computationally affordable and their data are simple to read, but the reliability of their data is low due to the environmental disturbances. The sonar range finder measures the distance to an object, but has poor angular resolution due to its wide beam width [3]. Laser range finders provide better reliability, instantaneous measurement, superior range accuracy, and precise angular resolution than sonar, with finer directional resolution, but at much higher cost. Laser-based sensors can extract information more than distance only. For example, laser scanners are often used to extract topological information making the best use of its ability to identify the textures of an object's surface and its precise range approximation. The laser range finder has a disadvantage that the scan may be prone to missing transparent objects, such as glasses and windows. Inertial navigation sensors such as accelerometers and gyroscopes provide orientation and trajectory measurements of the moving vehicles, but provide no information about the obstacles in the environment that the vehicle is traversing. GPS is one of the most popular aiding tools in navigation systems in use today. GPS provides real time absolute or relative position data, but the accuracy and bandwidth are limited compared to the typical requirements of relative proximity operation. The performance of GPS can suffer by occlusion of line-of-sight to satellites and their accuracy and update rate may be slow [4]. These range-based sensors have difficulties in detecting small or flat objects on the ground. These sensors are also unable to distinguish between difference types of ground surfaces. While small objects and different types of grounds are difficult to detect with range-based sensors, they can in many cases be easily detected with a passive sensor, like camera. A vision system is considered as a passive sensor and has the fundamental advantages over the active sensors that are considered as active sensors such as infrared, laser, and sonar sensors [5]. Passive sensors such as cameras do not alter the environment by emitting lights or waves in acquiring data, and also the obtained image contains more information (i.e. substantial, spatial and temporal information) than active sensors. All these sensors acquire less information about the physical environment than a camera can potentially, and with the continued growth of faster and cheaper computing power, that potential is now being tapped for designing real-world vision based navigational systems. Cameras are cheap to purchase, with even the most expensive cameras being relatively affordable. Hence vision as a sensing mechanism for mobile robots offers very attractive potential for solution.

1.3 Vision-Based Navigation

Vision based robot navigation is defined as the technique that guides a mobile robot to a desired destination, or along a desired path in an environment, by avoiding static (and may be dynamic) obstacles, primarily using vision sensor [100]. In general, vision-based robots have a vision system that perceives the external environments. Traditionally there are five main components in a vision system of an autonomous vision-based robot [6].

1. Maps. The system requires some internal representation or knowledge of the external environment in order to perform goal driven tasks.
2. Data Acquisition. The system collects images from a camera.
3. Feature Extraction. The feature extraction stage extracts significant features from input images such as edge, texture and colour.
4. Landmark Recognition. The system searches for possible matches between the features in the observed images and the expected landmarks pre-stored in memory with respect to some preset criteria.
5. Self-Localisation. The self-localisation stage calculates the robot's current position as a function of detected landmarks and its previous position. The system then derives the path for the robot to traverse. This traversal can be reactive to avoid obstacles only and/or it can be goal driven.

The navigation problem of a mobile robot can be, most often, divided into four subproblems [7]:

1. World perception. It senses the world, symbolizing it into features.
2. Path planning. It uses the features to create an ordered sequence of objective points that the robot must attain.
3. Path generation. Then the goal is to obtain a path through the sequence of objective points.
4. Path tracking. It is the responsibility of the controller so that the mobile robot can follow the intended path.

Generally, vision-based robotic systems with the ability of obstacle detection and avoidance are relatively complicated to develop, since extracting information from a stream of the images of the site, consisting of the robot and the obstacles, can be a complex task to achieve desired real-time performance, with as little computing processing as possible. The problem of moving a robot through an unknown environment has attracted much attention over the past two decades. A robot may encounter obstacles of all forms that must be bypassed in an intelligent manner. Accordingly, a substantial research effort focuses on the use of computer vision to achieve vision-based autonomous mobile robotic systems capable of navigation by logically acting on the sensed data to avoid such obstacles. The primary aim of most of these research efforts is to locate hindering obstacles, both stationary and movable, so that the suitable robot path can be planned to bypass these objects and, finally, to act according to the resultant plan. Navigation in both indoor and outdoor environments using vision sensing has evolved as two major research areas in the mobile robotics community.

1.3.1 Vision Based Indoor Navigation

The vision based indoor navigation schemes can be broadly classified into three groups: i) map-based navigation, ii) map-building based navigation, and iii) mapless navigation [6].

1.3.1.1 Map-Based Navigation

In map-based navigation, the system has an *a priori* knowledge about the environment and the navigation system works with the knowledge of this map. These environment maps are provided in form of geometric models, topological maps, or sequence of images [8, 9, 10]. Early methods containing these maps had several degrees of details regarding the environment and were provided in form of "occupancy map", "virtual force fields", or "S-map" [11]. These methods were very prone to sensor errors and this problem was later addressed in the works of [12]. They made the very important suggestion to consider a tolerance about the uncertainties in sensor measurement [12]. These methods were called "absolute localization" and these research problems were later modified to solve for "incremental localization". Here it is assumed that an approximate knowledge about the location of the robot is available and it is incrementally refined during navigation process, as observations are made with the vision tool, and necessary actions are taken for subsequent navigation. The FINALE [13] system is based on this concept where a geometrical representation of space and a statistical uncertainty model for the location of the robot is used and a Kalman-filter based approach is employed to update the mean and the covariance matrix of the robot position, when a landmark is matched with an image feature. Another class of approach was based on topological representation of space where one can employ bank of neural networks, as in NEURO-NAV [14], or a more sophisticated version of NEURO-NAV employing supervisory fuzzy controller, called FUZZY-NAV [15]. However, map-based navigation methods suffer from the disadvantage that it is not easy to generate a model or map, specially metric maps, of the environment.

1.3.1.2 Map-Building-Based Navigation

Map-building based navigation procedures attempt to take care of this problem where the robots start with no *a priori* information about the environment, they explore the environment at first and build an internal description and then they proceed with the task of navigation, using that internal description. Most of the works in this domain utilize approaches based on topological representation of space, e.g. [16] and [17], and address several issues like how to construct a node in a graph-based description of space, how to distinguish between several neighboring nodes, how to consider the effect of sensor uncertainty etc. However, a chief drawback of these methods is the difficulty in recognizing those nodes, which were previously visited. Other approaches in this domain include the works involving occupancy-grid-based representation [18, 19], works employing panoramic view [20], and works which try to incorporate both the good features of occupancy-grid-based and topology-based approaches [21]. There are some other types of map-building navigation systems also described in literature, for example visual sonar [22] or local map-based system [23]. These systems collect data of the environment as they navigate and build a local map that is used as a support for on-line safe navigation. This local map includes information about specific obstacles and free space data of a reduced portion of the environment, which is usually a function of the camera field-of-view.

1.3.1.3 Mapless Navigation

Mapless navigation approaches fall in an even more ambitious category, where the navigation process starts and continues without any map. These kind of navigation procedures may also be called reactive navigation where important information about the environment are extracted online through observation, feature or landmark identification (usually in form of natural objects like walls, desks, doorways, corners etc.) and feature tracking and the navigation algorithm takes its decision as a "reaction" to these relevant, meaningful information extracted. Some of the traditionally popular approaches among these techniques employ optical-flow based techniques [24] and appearance-based techniques [25], [26]. Optical-flow based methods mimic the visual behavior of bees where the motion of the robot is determined on the basis of the difference in velocity between the image seen with left eye of the robot and the image seen with the right eye of the robot. The robot moves toward the side whose image changes with smaller velocity. Further modifications of the basic method in [24] have shown development of optical flow based navigation systems utilizing depth information and also more sophisticated systems employing stereo heads with pan, tilt and vergence control. Some authors have proved that the combination of stereo vision, to obtain accurate depth information, and optical flow analysis, provides better navigation results [27, 28]. In [29], stereo information is combined with the optical flow from one of the stereo images, to build an occupancy grid and perform a real time navigation strategy for ground vehicles. On the other hand, appearance-based methods thrive on memorizing the environment by storing a series of images, usually created from subwindows extracted from down-sampled camera images, and then, at any given time, an image taken is scanned across all these templates to find out whether the image matches with any of the stored ones. If the match is found, then a corresponding control action is taken for suitable navigation. The main focus here is on improving the way the images are recorded in the training phase, as well as on the subsequent image matching process. There are two main approaches for environment recognition without using a map [30]:

(i)   model based approaches which utilize pre-defined object models to recognize features in complicated environments and self-localize in it, and

(ii)   view-based approaches where no features are extracted from the pre-recorded images. The self-localization is performed using image matching algorithms.

1.3.2 Vision Based Outdoor Navigation

These are systems that use no explicit representation at all about the space in which navigation is to take place, but rather resort to recognizing objects found in the environment or to tracking those objects by generating motions based on visual observations. The outdoor navigation fall in two sub groups based on the level of structure of the environment: (i) outdoor navigation in structured environment, and (ii) outdoor navigation in unstructured environment. In many cases, mapping representations adapted in the indoor navigation are not much reliable for outdoor navigation, as they include large scenarios, with enormous physical area, and hence the amount of information to represent the environment increases immeasurably. The outdoor navigation in structured environments refers to road following which has the ability to detect the lines of the road and navigate consistently. In contrast with the structured indoor spaces, outdoors are in most cases, composed of gravel, gardens, walkways and streets. Most of these elements present different colors and textures and it is convenient to use these features for outdoor navigation. The first step to identify navigation regions is classifying portions of the terrains into classes, according to the visual information. One of the most outstanding efforts in road following, reported till now, is the NAVLAB project [31, 32]. The NAVLAB road following algorithm has three phases: in the first phase, a combination of color and texture pixel classification is performed by utilizing a Gaussian distribution for each road and non-road pixels; in the second phase, a Hough transform and a subsequent voting process is applied to road pixels, to obtain the road vanishing point and orientation parameters; finally, pixels are classified again according to the determined road edges. This classification procedure is repeated for the next image in order to have a system adaptable to changing road conditions. Many works have been reported with related concept for road detection and following in structured environment [33-36]. Supporting vision information with GPS data in outdoor environments is another possibility of increasing reliability in position estimation [37]. In recent works, authors have also proposed to combine the concept of feature tracking with stereo 3D environment reconstruction. In [38], stereo vision is used in a novel navigation strategy applicable to unstructured outdoor environments. This system is based on a new, faster and more accurate corner detector method. In this method, detected features are 3D positioned and tracked using normalized mean-squared differences and correlation measurements.

1.4 State of the Art

Numerous research works have been conducted in the field of vision based mobile robotics, till now, and most studies are concerned with detecting obstacles, mapping a surrounding environment, planning safe routes, and navigating a doorway. In this section some of the different navigation approaches that have so far been used in vision based navigation, with and without *a priori* environmental information, based on visual information provided by the camera, are outlined.

The "Stanford Cart" is one of the well known vision-based mobile robot projects [39], implemented quite some time back. The system utilized a video camera mounted on a mobile platform, the video signal was broadcast to a remote computer which then processed and controlled the motion of the robot via radio signals. The system used a planner for obstacle avoidance and path determination. Once images of the complete scene of the environment were received, the system could process them. The processing time was reported to be as long as 1-2 hours. Once the image processing was completed, a planned path was produced which guided the robot around obstacles. The problems encountered by this system were largely due to the length of time the system spent processing the information. For example, the shadows moved, causing the robot to make errors in its maps. These drawbacks were later addressed, by incorporating the idea of using image segmentation, using an interest-operator, to detect distinctive features in an environment [40].

In early 1990's, Horswill developed a robot called POLLY [41, 42], which navigated using monochrome vision and was operated in a restricted environment with constant color. POLLY used a topology-like map in its navigation, which comprised a set of landmarks, that it used to localise itself. The landmarks were a set of individual snapshots taken from a particular location in the office environment. Landmark or place recognition was carried out by matching every landmark with live video data, in order to determine where the robot was currently located. In this work, it was suggested that this was, in effect, a local navigation strategy, equivalent to the method of artificial potential fields. Therefore, the robot was prone to be trapped in local minima. The approach utilized in this work mainly involved four steps: (i) smooth the image, (ii) Determine the average pixel value from a foreground trapezium, (iii) use this as floor and then label every pixel by starting from the bottom of the image and scanning up each column until there is a mismatch, (iv) the height of columns indicates the distance to obstacles, referred as the Radial Depth Map. The main problem with this methodology was that at times it misinterpreted shadows as obstacles. This was mainly due to the simple nature of the robot's vision processing system. The processing was based on a simple extraction of textureless floor in an image, determining the available free space to travel. In later years, researchers have also proposed mobile robot navigation schemes with only one camera [43, 44]. Research work employing occupancy gird based map building framework and a feature position detection algorithm, that processes the colour RGB image sequence on-line from a single camera, has also been proposed [45]. This system, instead of implementing matching approaches, computes probabilities of finding objects at every location. The algorithm starts with detecting the edges of objects boundaries in the current frame using the Harris edge and corner detectors. In the beginning of the image sequence, the edge features are added to the occupancy grid map, which are scanned to determine the peaks. The detected features are back projected from the

2D image plane, considering all the potential locations at any depth. The positioning module of the system computes the position of the robot using odometry data combined with image feature extraction. Color or gradient from edges and features from past images help to increase the confidence one can have in the presence of an object in a certain location. The size of the grid cells was set to 25 × 25 mm for experiments carried out in indoor environments. The robot was allowed to move 100 mm between consecutive images. Using a single camera, only forward information was acquired, and this amount of information was found sufficient for indoor navigation purpose and was found very cost effective. Majority of the other works using single camera, for indoor unknown environment, have focused on the center following and wall following methods [46]. A corridor center method for wheel chair mobile robot navigation using a single USB camera and a laptop is reported recently [47]. In this work, the size of the acquired image is 320 x 240 pixels, the frontal field-of-view is 60 cm at the bottom and 20 m at the top, with a moving speed of 0.823 Km/h. This method also used Hough transform to detect the boundary lines of the corridor and the walls. The robot moves at the center of the corridor when there is no obstacles in the corridor. The obstacle detection used here is based on an improved version of Ulrich's method [95]. The improvements incorporated are to omit the false detection of obstacles, caused by the influence of the lighting. In [47], if any obstacle is detected, then the obstacle avoidance or stop movement is decided based on the size of the obstacle, distance of the mobile robot from the obstacle, and the width of the corridor, which is determined from the 2D position in the real space and the arbitrary position in the image. Another work has been reported based on qualitative approach, which uses a single off-the-shelf, forward looking camera with no calibration, which can perform both indoor and outdoor navigation [1]. In this work the approach is based on teach-replay method, where, during the teaching phase, a human guides the robot along the path which it should traverse, manually. During the teaching phase, the robot selects and tracks the feature points using Kanade-Lucas-Tomasi (KLT) feature tracker [48] and stores then in a database. During the replay phase, an attempt is made to establish a correspondence between the feature point coordinates of the current image, with those of the first image taken during the teaching phase, based on which the turning commands are determined. A similar type of human experience based navigation algorithm using teaching-replay technique has also been developed using stereo vision [49].

In stereo vision, one can measure distance information with two or more cameras, as well as using ultrasonic sensor. However the processing cost becomes complex when two or more cameras are used. Several works in mobile robot navigation have so far been reported by using two or more cameras [50-54]. To obtain the depth information by the use of two cameras, it is necessary to have some data about the geometry of the camera and the head used. To obtain depth information in stereo vision, it is required that the two lines of sight for the two cameras intersect at the scene point P for which the depth information is to be processed. Stereo camera based systems are useful for feature identification, tracking of features and the distance calculation of 3D feature points in real time, for the purpose of navigation. The concept relies heavily on selection/extraction of image points/features from a snap acquired and subsequently tracking of it (them) in subsequent snaps for the same scene, acquired from a different location and/or with a different orientation. No feature based vision system can work unless good features can be identified and tracked [55]. It is very important to obtain the 3D coordinates of image features which can facilitate the calculation of distance between the selected feature(s) in the image plane and the focal point of the camera in the field of robotics. This is very important for robot localization and scene interpretation. Recently, some methods to measure the distance between the feature or object and a camera have been developed using a fisheye stereo vision [56], monocular camera [57], the integration of vision and ultrasonic sensors [58], biologically inspired saliency Maps (SMs) [59] that receive preprocessed input from feature Detectors (FDs) etc. In [59], the interaction between the FDs of both cameras and SMs support the detection of corresponding landmarks in both images and allow the estimation of their direction and distance. A neural network maps the seven given identifiers (the X and Y positions of the landmark in both images, two camera pan angles and one common camera tilt angle) to the direction and distance of the landmark. A grid-based map building method, by using stereo vision, was developed for LAGR robot for outdoor navigation [60]. In addition to the vision sensor, inertial navigation unit, GPS receiver, and front bump switch were incorporated for sensing purpose. Here the map-building method mainly involves the following steps:

1. Grab the images using a pair of color camera.
2. 3D representation is determined by matching patches in the two images from the relative geometry of the camera.
3. Each coordinate point is transformed to find the instantaneous pitch and roll of the robot, as estimated by the robots inertial navigation and yaw from the local frame into the global frame.
4. A derivative operation is applied to the terrain map to find the abrupt changes in the slope.
5. The global map is updated with new measurement including the terrain and derivative estimates.

Recently, a successful stereo vision based algorithm was developed for NASA's Mars Exploration Rover, for autonomous navigation in potentially hazards terrain [61]. The processing steps involved in this project are: (i) the image received from the camera is down sampled to 256 x 256 pixels for reducing the computation time, (ii) then the pair of image is processed by projection of epipolar line of an object in the first image to the same object on the other image horizontally, (iii) then the laplacian of the two images were computed, and, were correlated to select the potential match within a disparity range and the procedure was repeated for all the pixels in the images, so that if the estimate of these pixels fails in matching, it will be discarded, and (iv) finally, each disparity value can be mapped to a 3D representation using the geometric camera model.

Utilization of Omni-directional vision is also an important part in developing vision-based mobile robot navigation strategies. It can provide a 360^0 view of the environment, in image form, for any arbitrary position of the robot. This vision system uses fish eye cameras and panoramic cameras. They have the advantage of possessing the full field-of-view but the associated disadvantage of incurring high cost and the complexity in the development of vision algorithms, based on the geometry of the particular type of the camera chosen. Several works have also been reported using omni-directional vision for navigation [62-65], mostly using the optical flow based navigation techniques.

Optical flow is the measure of visual motion induced by the movement of surfaces in a scene with respect to the camera. Computationally, the most common representation of optical flow is a 2D vector field in the image space, where each vector describes the motion of a point in one image to its location in the next. In a vision-guided robot, optical flow is largely induced by the motion of the camera as the robot moves. Many of the works in optical flow methods are corridor centering approaches, based on the observation in flying honey bees, as mentioned earlier. It was observed that the direction and speed of the flight of the honey bees is directly coupled with the visual motion induced by its motion relative to the environment [66, 67]. The corridor navigation of mobile robots, based on optical flow, is achieved using different types of cameras with different placement. For example, in [68], a wide angle, forward-facing active camera was used to achieve corridor centering. Here optical flow was computed in the left and right peripheral thirds of the image. By balancing the maximal flow in both thirds, the robot attempted to maintain a centered path between walls and obstacles. Here the camera gaze direction was used to counter the rotation of the robot during directional adjustments. In [69, 70], different methodologies were utilized to compute the difference of the average horizontal optical flow from two cameras placed at 90^0 to the heading direction, on either side of the robot. In this method, in contrast to [68], no compensatory measures were taken to counter rotational flow. Instead, restrictions on steering keep the induced rotational effects to a minimum. To cope with absence of texture (e.g. a doorway or no wall), a unilateral sustaining behavior is used to maintain a constant distance from the side wall, which can still provide sufficient texture from which optical flow can be estimated.

Optical flow using correlation-based techniques is similar in nature to disparity mapping using stereo vision. The difference is that in optical flow the images are separated temporally, whereas, for stereo vision, they are separated spatially. Unfortunately, the task of correlating images for optical flow gets complicated by the fact that robots in real world are may be subject to vibrations. This means that, unlike stereo vision, the search for corresponding image patches cannot be restricted to the same horizontal scan line [71, 72]. Other flow techniques for robot navigation with continuous motion have also been proposed using gradient-based methods [73-75].

Visual navigation techniques based on optical flow have proved to be especially useful for unmanned aerial vehicles because optical flow gives the scene qualitative characteristics that cannot be extracted in detail from a single low quality image. Within this research framework, a significant effort has been devoted to imitate animal behavior, as far as the use and processing of apparent motion is concerned. Unmanned aerial vehicles with camera eye consisted of an array of photoreceptors, each one connected to an electronic Elementary Motion Detector (EMD), which was able to calculate the local optical flow at its particular position [76]. Contrast on optical flow calculations determined the presence of obstacles, while identifying the EMD polar coordinates, that produced the changes in optical flow measures, permitted to construct a local map with the location of the obstacles. In [77], an unmanned aerial vehicle was also implemented with a camera eye, assembled with an array of photosensors and their corresponding EMDs. The information obtained from the set of EMDs, was used to determine the presence of obstacles. Furthermore, when the unmanned aerial vehicle flew at a constant speed and altitude, a reference optical flow distribution was calculated from the equation that models the velocity of the artificial retina. To follow the terrain, the system varied thrust and rudders positions to adjust the online computed optical flow with the optical flow reference.

Line navigation is another type of landmark based navigation that has been widely used in the industries. Line navigation can be thought of as a continuous following of a landmark, although in most cases the sensor used in this system needs to be very close to the line, so that the range of the vehicle is limited to the immediate vicinity of the line. These techniques can be popularly employed for industrial automation tasks and vehicles using them are generally called Automatic Guided Vehicles (AGVs). An automated guided vehicle can navigate in prespecified paths where the work is monotonous such as in factory, hospital, and office building [78]. In earlier times, cable magnetic tape guidance was the preferred choice for line navigation. But the main disadvantage of this method was the cost involved and the difficulty in relocating the paths which leads navigation using line recognition. One of the successful early approaches of vision based line following navigations is based on image processing by extracting a white line from the background of the image acquired and it varies with respect to the vehicles movement [79]. In this work, a TV camera was used for acquiring the environmental information. The position of the vehicle is determined by correlating the field pattern changes while in movement with the predetermined path in the path planner accordingly providing the steering command to the robot motor. Similarly vehicle navigation based on preexisting landmarks with signs and lines were reported in [80, 81].

Another area, which has attracted significant research attention, is simultaneous localization and mapping (SLAM), also known as concurrent mapping and localization (CML), where a mobile robot can build a map of an environment and at the same time use this map to deduce its location. Initially, both the map and the vehicle position are not known, the vehicle has a known kinematic model and it is moving through the unknown environment, which is populated with artificial or natural landmarks. A simultaneous estimation of both robot and landmark locations is carried out based on observation of landmarks. The SLAM problem involves finding appropriate representation for both the observation and the motion models [82]. Most of the SLAM approaches are oriented towards indoor, well structured and static environment [83-87] and give metric information regarding the position of the mobile robot and of the landmarks. A few works have also been attempted for dynamic scenarios and for outdoor environments [88, 89]. In the earlier stages of mapping algorithms using sonar and vision sensors on large experimental areas, it was noticed that there were storage problems for specifically e.g. long straight walls. As a mobile robot moves, errors in the odometry information arising from wheel slippage, non-uniform floor surface, and poorly calibrated tick-information causes the position information provided by the odometry to increasingly deviate from its true position. Features detected from these positions would be built into the map relative to the position of the robot, hence the positions of features would also drift away from their true positions. Algorithms have been developed to correct for this motion drift such as [90] and [91] who proposed to store correlations between each feature and robot position. One of the first vision based solution for SLAM problems was proposed in [90] which employed an extended Kalman filter (EKF)-based approach. Although EKF-based approaches are more common for these problems, they are based on the basic assumption that the sensor and process uncertainties can be modeled as Gaussian distributions. However, physical systems can have significant departure from these assumptions. One of the main drawbacks of the EKF and the KF implementations is the fact that for long duration missions, the number of landmarks will increase and, eventually, computer resources will not be sufficient to update the map in real-time. This scaling problem arises because each landmark is correlated to all other landmarks. The Compressed Extended Kalman Filter (CEKF) [87] algorithm significantly reduces the computational requirement without introducing any penalties in the accuracy of the results. A CEKF stores and maintains all the information gathered in a local area with a cost proportional to the square of the number of landmarks in the area. This information is then transferred to the rest of the global map with a cost that is similar to full SLAM but in only one iteration. To overcome these problems, recently some efforts in the area of vision-based SLAM are directed in utilizing particle filtering based approaches [92, 93]. However, particle filtering is essentially a slow process and hence its real-time implementation can cause significant problems.

1.5 Obstacle Detection and Avoidance

Obstacle avoidance is one of the important steps in the role of most mobile robot navigation, schemes. Obstacle detection is the process of discriminating between the floor (also called the ground plane) and an object resting on the floor, i.e. separating the floor pixels from the obstacle pixels in the camera image. Collision avoidance is a steering behavior that enables a robot to roam around without colliding with obstacles. As the present book is based on vision-based navigation, the discussions here will be restricted to those works where the steering decisions are made based on computer vision. Many works described before have the ability of obstacle avoidance with the support of other conventional sensors. There are also several methods adapted for obstacle avoidance with monocular image features, stereo vision based detection, optical flow method, and vision based potential field method. In [94], an autonomous obstacle detection method for mobile robots, using single monocular camera image, has been proposed. This system basically comprises three vision modules for obstacle detection. The three modules of vision processing are based on brightness gradients, RGB, and HSV and they generate a coarse image-based representation, called obstacle boundary. The outputs of these three modules were combined into a single obstacle boundary and this information is utilized to generate the turning commands. The purpose of utilizing three modules is that at any circumstances, two of the modules will be suitable for detecting the boundary. This method has been tested for two simulated Mars environments at JPL (Jet Propulsion Laboratory). However, the disadvantage of this system is that it failed when the obstacles were outside the field-of- view of the camera.

Another strategy for obstacle avoidance is based on appearance based method for structured environments [95]. This system is based on three assumptions: (i) the obstacles differ in appearance from the ground, (ii) the ground is relatively flat, and (iii) all the obstacles should be in touch with the ground. The system uses an image resolution of 320 x 260 pixels in colour. The main process comprises four steps:

1. Filter the colour image using a 5 x 5 gaussian, mask.
2. Transform the filtered colour image to HIS colour space.
3. The pixels inside the trapezoidal area are histogrammed for hue and intensity.
4. All pixels of the filtered image are compared to the hue and intensity histograms. If the histogram's bin value at the pixel's hue and pixel's intensity value is below the threshold, then it is classified as obstacle.

This system has three operating modes: regular, adaptive and assistive. Each of these modes is well suited for a specific situation. In adaptive mode, the system can cope up with the changes in the illumination and in the assistive mode, the robot is equipped for tele operation.

Another method of obstacle detection and avoidance, by using visual sonar, is proposed in [96]. In this method, a single camera is mounted on the robot and each camera image pixel is classified into floor pixels, other known objects or unknown objects, based on their colour classes. The image is scanned with reference to the robot with a linear spacing of 5^0. An object is identified if there exists a continuous set of pixels in a scan which corresponds to the same colour class. The unknown obstacle is detected when unknown colour classes occur together. The distance between the robot and the image is calculated using the difference in the colour class value at the nearest intersecting pixel point of the object and the floor colour. A local map is created with the distance of objects from the robot, with the limited field-of-view of the camera. The new visual information is updated every time a new object information appears in the field-of-view of the camera. This vision algorithm is implemented in the AIBO robot. Another method of obstacle detection and avoidance with the combination of single camera and ultrasonic sensor is reported in [97]. In this work, the obstacle detection is carried out using canny edge detection method in vision, and the obstacle avoidance is carried out using limit-cycle and nearness diagram navigation method.

In stereo vision based obstacle detection methods, the main idea lies in capturing two images of the environment at the same time. The position of an obstacle can be determined by inverse perspective mapping [98]. In inverse perspective mapping, the pixels of the two images are mapped to the ground plane, as if they all represented points on the ground. The obstacle positions are calculated using the difference of two images, because the difference signifies the presence of an obstacle. The drawback of this method is that it is essentially a computation heavy procedure. Obstacle detection and avoidance for outdoor environments, based on the computation of disparity from the two images of a stereo pair of calibrated cameras, was also reported in [99]. In this work, the system assumed that objects protrude high from the ground, and the surface should be flat, distinguishable from the background in the intensity image. Every point above the ground is configured as a potential object and projected onto the ground plane, in a local occupancy grid, called instantaneous obstacle map (IOM). The commands to steer the robot are generated according to the positions of obstacles in this instantaneous obstacle map computed.

1.6 Summary

This chapter has introduced the fundamental concepts of autonomous mobile robot navigation using vision. Different broad categories of vision-based navigation are discussed and the research efforts worldwide, in the present day context, in these fields, are summarized.

References

[1]   Chen, Z., Birchfield, S.T.: Qualitative Vision-Based Mobile Robot Navigation. In: Proc. IEEE International Conference on Robotics and Automation (ICRA), Orlando, Florida (May 2006)

[2]   Benet, G., Blanes, F., Simo, J.E., Perez, P.: Using infrared sensors for distance measurement in mobile robots. Robotics and Autonomous Systems 40, 255–266 (2002)

[3]   Flynn, A.M.: Combining sonar and infrared sensors for mobile robot navigation. The International Journal of Robotics Research 7(6), 5–14 (1988)

[4]   Saeedi, P., Lawrence, P.D., Lowe, D.G., Jacobsen, P., Kusalovic, D., Ardron, K., Sorensen, P.H.: An autonomous excavator with vision-based track-slippage. IEEE Transaction on Control Systems Technology 13(1), 67–84 (2005)

[5]   Bertozzi, M., Broggi, A., Fascioli, A.: Vision-based intelligent vehicles: state of the art and perspectives. Robotics and Autonomous Systems 32, 1–16 (2000)

[6]   DeSouza, G.N., Kak, A.C.: Vision for mobile robot navigation: A Survey. IEEE Transactions on Pattern Analysis and Machine Intelligence 24(2), 237–267 (2002)

[7]   Shin, D.H., Singh, S.: Path generation for robot vehicles using composite clothoid segments. The Robotics Institute, Internal Report CMU-RI-TR-90-31. Carnegie-Mellon University (1990)

[8]   Lebegue, X., Aggarwal, J.K.: Generation of architectural CAD models using a mobile robot. In: Proc. IEEE International Conference on Robotics and Automation (ICRA), pp. 711–717 (1994)

[9]   Lebegue, X., Aggarwal, J.K.: Significant line segments for an indoor mobile robot. IEEE Transactions on Robotics and Automation 9(6), 801–815 (1993)

[10] Egido, V., Barber, R., Boada, M.J.L., Salichs, M.A.: Self-generation by a mobile robot of topological maps of corridors. In: Proc. IEEE International Conference on Robotics and Automation (ICRA), Washington, pp. 2662–2667 (May 2002)

[11] Borenstein, J., Everett, H.R., Feng, L. (eds.): Navigating Mobile Robots: Systems and Techniques. A. K. Peters, Wellesley (1996)

[12] Atiya, S., Hanger, G.D.: Real-time vision based robot localization. IEEE Transactions on Robotics and Automation 9(6), 785–800 (1993)

[13] Kosaka, A., Kak, A.C.: Fast vision-guided mobile robot navigation using model-based reasoning and prediction of uncertainties. Computer Vision, Graphics, and Image Processing – Image Understanding 56(3), 271–329 (1992)

[14] Meng, M., Kak, A.C.: Mobile robot navigation using neural networks and nonmetrical environment models. IEEE Control Systems, 30–39 (October 1993)

[15] Pan, J., Pack, D.J., Kosaka, A., Kak, A.C.: FUZZY-NAV: A vision-based robot navigation architecture using fuzzy inference for uncertainty. In: Proc. IEEE World Congress Neural Networks, vol. 2, pp. 602–607 (July 1995)

[16] Yamauchi, B., Beer, R.: Spatial learning for navigation in dynamic environments. IEEE Transactions on Systems, Man, and Cybernetics: Part B 26(3), 496–505 (1996)

[17] Zimmer, U.R.: Robust world-modeling and navigation in real world. In: Proc. Third International Conference Fuzzy Logic, Neural Nets, and Soft Computing, vol. 13(2-4), pp. 247–260 (October 1996)

[18] Borenstein, J., Koren, Y.: The vector-field histogram-fast obstacle avoidance for mobile robots. IEEE Transactions on Robotics and Automation 7(3), 278–288 (1991)

[19] Elfes, A.: Sonar-based real-world mapping and navigation. IEEE Journal of Robotics and Automation 3(6), 249–265 (1987)

[20] Yagi, Y., Kawato, S., Tsuji, S.: Real-time ominidirectional image sensor (COPIS) for vision guided navigation. IEEE Transactions on Robotics and Automation 10(1), 11–22 (1994)

[21] Thrun, S.: Learning metric-topological maps for indoor mobile robot navigation. Artificial Intelligence 99(1), 21–71 (1998)

[22] Martin, M.C.: Evolving visual sonar: Depth from monocular images. Pattern Recognition Letters 27(11), 1174–1180 (2006)

[23] Gartshore, R., Palmer, P.: Exploration of an unknown 2D environment using a view improvement strategy. Towards Autonomous Robotic Systems, 57–64 (2005)

[24] Santos-victor, J., Sandini, G., Curotto, F., Garibaldi, S.: Divergent stereo for robot navigation: learning from bees. In: Proc. IEEE CS Conference Computer Vision and Pattern Recognition (1993)

[25] Ohno, T., Ohya, A., Yuta, S.: Autonomous navigation for mobile robots referring pre-recorded image sequence. In: Proc. IEEE International Conference on Intelligent Robots and Systems, vol. 2, pp. 672–679 (November 1996)

[26]   Jones, A.D., Andersen, C., Crowley, J.L.: Appearance based processes for visual navigation. In: Proc. IEEE International Conference on Intelligent Robots and Systems, pp. 551–557 (September 1997)

[27]   Talukder, A., Goldberg, S., Matties, L., Ansar, A.: Real-time detection of moving objects in a dynamic scene from moving robotic vehicles. In: Proc. IEEE International Conference on Intelligent Robots and Systems (IROS), Las Vegas, Nevada, pp. 1308–1313 (October 2003)

[28]   Talukder, A., Matties, L.: Real-time detection of moving objects from moving vehicles using dense stereo and optical flow. In: Proc. IEEE International Conference on Intelligent Robots and Systems (IROS), Sendai, pp. 3718–3725 (October 2004)

[29]   Braillon, C., Usher, K., Pradalier, C., Crowley, J.L., Laugier, C.: Fusion of stereo and optical flow data using occupancy grid. In: Proc. IEEE International Conference on Intelligent Robots and Systems (IROS), Beijing, pp. 2302–2307 (October 2006)

[30]   Matsumoto, Y., Ikeda, K., Inaba, M., Inoue, H.: Visual navigation using omnidirectional view sequence. In: Proc. IEEE International Conference on Intelligent Robots and Systems (IROS), Kyongju, Korea, pp. 317–322 (October 1999)

[31]   Thorpe, C., Herbert, M.H., Kanade, T., Shafer, S.A.: Vision and Navigation for the Carnegie-Mellon Navlab. IEEE Transactions on Pattern Analysis and Machine Intelligence 10(3), 362–372 (1988)

[32]   Thorpe, C., Kanade, T., Shafer, S.A.: Vision and Navigation for the Carnegie-Mellon Navlab. In: Proc. Image Understand Workshop, pp. 143–152 (1987)

[33]   Broggi, A., Berte, S.: Vision-based road detection in automotive systems: A real-time expectation-driven approach. Journal of Artificial Intelligence Research 3(6), 325–348 (1995)

[34]   Ghurchian, R., Takahashi, T., Wang, Z.D., Nakano, E.: On robot self navigation in outdoor environments by color image processing. In: Proc. International Conference on Control, Automation, Robotics and Vision, pp. 625–630 (2002)

[35]   Jung, C.R., Kelber, C.R.: Lane following and lane departure using a linear-parabolic model. Image and Vision Computing 23(13), 1192–1202 (2005)

[36]   Schneiderman, H., Nashman, M.: A discriminating feature tracker for vision-based autonomous driving. IEEE Transactions on Robotics and Automation 10(6), 769–775 (1994)

[37]   Mejias, L.O., Saripalli, S., Sukhatme, G.S., Cervera, P.C.: Detection and tracking of external features in an urban environment using an autonomous helicopter. In: Proc. IEEE International Conference on Robotics and Automation (ICRA), Barcelona, pp. 3972–3977 (April 2005)

[38]   Saeedi, P., Lawrence, P.D., Lowe, D.G.: Vision-based 3-D trajectory tracking for unknown environments. IEEE Transaction on Robotics 22(1), 119–136 (2006)

[39]   Moravec, H.P.: The stanford cart and the CMU rover. Proc. IEEE 71(7), 872–884 (1983)

[40]   Thorpe, C.: FIDO: Vision and navigation for a mobile robot. PhD dissertation, Department of computer science, Carnegie Mellon University (December 1984)

[41]   Horswill, I.: Visual collision avoidance by segmentation. In: Proc. IEEE International Conference on Intelligent Robots and Systems, Germany, pp. 902–909 (September 1994)

[42] Horswill, I.: Specialzation of Perceptual Processes. PhD thesis, Massachusetts Institute of Technology (1995)

[43] Ohya, A., Kosaka, A., Kak, A.: Vision-based navigation by a mobile robot with obstacle avoidance using single-camera vision and ultrasonic sensing. IEEE Transactions on Robotics and Automation 14(6), 969–978 (1998)

[44] Aider, O.A., Hoppenot, P., Colle, E.: A model-based method for indoor mobile robot localization using monocular vision and straight-line correspondences. Robotics and Autonomous Systems 52, 229–246 (2005)

[45] Gartshore, R., Aguado, A., Galambos, C.: Incremental map building using occupancy grid for an autonomous monocular robot. In: Proc. Seventh International Conference on Control, Automation, Robotics and Vision (ICARCV), Singapore, pp. 613–618 (December 2002)

[46] Murillo, A.C., Kosecka, J., Guerrero, J.J., Sagues, C.: Visual door detection integrating appearance and shape cues. Robotics and Autonomous Systems 56, 512–521 (2008)

[47] Saitoh, T., Tada, N., Konishi, R.: Indoor mobile robot navigation by center following based on monocular vision. In: Computer Vision, pp. 352–366. In-teh Publishers

[48] Birchfield, S.: KLT: An implementation of the Kanade- Lucas-Tomasi feature tracker, http://www.ces.clemson.edu/~stb/klt/

[49] Kidono, K., Miura, J., Shirai, Y.: Autonomous visual navigation of a mobile robot using a human guided experience. Robotics and Autonomous Systems 40(23), 124 132 (2002)

[50] Murray, D., Little, J.J.: Using real-time stereo vision for mobile robot navigation. Autonomous Robots 8, 161–171 (2000)

[51] Davison, A.J.: Mobile robot navigation using active vision. PhD thesis (1998)

[52] Ayache, N., Faugeras, O.D.: Maintaining representations of the environment of a mobile robot. IEEE Transactions on Robotics and Automation 5(6), 804–819 (1989)

[53] Olson, C.F., Matthies, L.H., Schoppers, M., Maimone, M.W.: Rover navigation using stereo ego-motion. Robotics and Autonomous Systems 43(4), 215–229 (2003)

[54] Konolige, K., Agrawal, M., Bolles, R.C., Cowan, C., Fischler, M., Gerkey, B.: Outdoor Mapping and Navigation using Stereo Vision. In: Proc. International Symposium on Experimental Robotics (ISER), Brazil, pp. 1–12 (July 2006)

[55] Shi, J., Tomasi, C.: Good Features to Track. In: Proc. IEEE Conference on Computer Vision and Pattern Recognition (CVPR 1994), Seattle, pp. 593–600 (June 1994)

[56] Nishimoto, T., Yamaguchi, J.: Three dimensional measurements using fisheye stereo vision. In: Proc. SICE Annual Conference, Japan, pp. 2008–2012 (September 2007)

[57] Yamaguti, N., Oe, S., Terada, K.: A Method of distance measurement by using monocular camera. In: Proc. SICE Annual Conference, Japan, pp. 1255–1260 (July 1997)

[58] Chou, T.N., Wykes, C.: An integrated ultrasonic system for detection, recognition and measurement. Measurement 26, 179–190 (1999)

[59] Conradt, J., Simon, P., Pescatore, M., Verschure, P.F.M.J.: Saliency Maps Operating on Stereo Images Detect Landmarks and Their Distance. In: Dorronsoro, J.R. (ed.) ICANN 2002. LNCS, vol. 2415, pp. 795–800. Springer, Heidelberg (2002)

[60] Wooden, D.: A guide to vision-based map-building. IEEE Robotics and Automation Magazine, 94–98 (June 2006)

[61] Goldberg, S.B., Maimone, M.W., Matthies, L.: Stereo vision and rover navigation software for planetary exploration. In: Proc. IEEE Aerospace Conference Proceedings, USA, vol. 5, pp. 5025–5036 (March 2002)

[62] Fialaa, M., Basub, A.: Robot navigation using panoramic tracking. Pattern Recognition 37, 2195–2215 (2004)

[63] Gasper, J., Santos- Victor, J.: Vision-based navigation and environmental representations with an omnidirectional camera. IEEE Transactions on Robotics and Automation 16(6), 890–898 (2000)

[64] Winters, N., Santos-victor, J.: Ominidirectional visual navigation. In: Proc. IEEE International Symposium on Intelligent Robotic Systems (SIRS), pp. 109–118 (1999)

[65] Gasper, J., Winters, N., Santos-victor, N.: Vision-based navigation and environmental representation with an ominidirectional camera. IEEE Transtations on Robotics and Automation 16(6), 890–898 (2000)

[66] Srinivasan, M.V.: An image-interpolation technique for the computation of optic flow and Egomotion. Biological Cybernetics 71(5), 401–415 (1994)

[67] Srinivasan, M.V., Zhang, S.: Visual navigation in flying insects. International Review of Neurobiology 44, 67–92 (2000)

[68] Coombs, D., Roberts, K.: Centering behaviour using peripheral vision. In: Proc. IEEE Computer Society Conference on Computer Vision and Pattern Recognition, USA, pp. 440–445 (June 1993)

[69] Sandini, G., Santos-Victor, J., Curotto, F., Garibaldi, S.: Robotic bees. In: Proc. IEEE/RSJ International Conference on Intelligent Robots and Systems, Yokohama, Japan, vol. 1, pp. 629–635 (1993)

[70] Santos-Victor, J., Sandini, G.: Divergent stereo in autonomous navigation: From bees to robots. International Journal of Computer Vision 14(2), 159–177 (1995)

[71] Lourakis, M.I.A., Orphanoudakis, S.C.: Visual Detection of Obstacles Assuming a Locally Planar Ground. In: Chin, R., Pong, T.-C. (eds.) ACCV 1998. LNCS, vol. 1352, pp. 527–534. Springer, Heidelberg (1997)

[72] Camus, T.: Real-time quantized optical flow. Real-Time Imaging 3(2), 71–86 (1997)

[73] Lucas, B., Kanade, T.: An iterative image registration technique with an application to stereo vision. In: Proc. DARPA Image Understanding Workshop, pp. 121–130 (1984)

[74] Horn, B.K.P., Schunck, B.G.: Determining optical flow. Artificial Intelligence 13, 185–203 (1981)

[75] Nagel, H.: On the estimation of optical flow: relations between different approaches and some new results. Artificial Intelligence 33(3), 299–324 (1987)

[76] van der Zwaan, S., Santos-Victor, J.: An insect inspired visual sensor for the autonomous navigation of a mobile robot. In: Proc. Seventh International Sysposium on Intelligent Robotic Systems, Portugal (July 1999)

[77] Netter, T., Franceschini, N.: A robotic aircraft that follows terrain using a neuromorphic eye. In: Proc. IEEE International Conference on Intelligent Robots and Systems (IROS), Switzerland, vol. 1, pp. 129–134 (Septemper 2002)

[78] Zhang, H., Yuan, K., Mei, S., Zhou, Q.: Visual navigation of automated guided vehicle based on path recognition. In: Proc. Third International Conference on Machine Learning and Cybernectics, Shanghai, pp. 26–29 (August 2004)

[79] Ishikawa, S., Kuwamoto, H., Ozawa, S.: Visual navigation of an autonomous vehicle using white line recognition. IEEE Transactions on Pattern Analysis and Machine Intelligence 10(5), 743–749 (1988)

[80] Beccari, G., Caselli, S., Zanichelli, F., Calafiore, A.: Vision-based line tracking and navigation in structured environments. In: Proc. IEEE International Symposium on Computational Intelligent in Robotics and Automation, USA, pp. 406–411 (July 1997)

[81] Ismail, A.H., Ramli, H.R., Ahmad, M.H., Marhaban, M.H.: Vision-based system for line following mobile robot. In: Proc. IEEE Symposium on Industrial Electronics and Applications (ISIEA), Malaysia, pp. 642–645 (October 2009)

[82] Durrant-White, H., Bailey, T.: Simultaneous localization and mapping. IEEE Robotics and Automation Magazine 13(2), 99–108 (2006)

[83] Zunino, G., Christensen, H.I.: Simultaneous localization and mapping in domestic environments. Multisensor Fusion and Integration for Intelligent Systems, 67–72 (2001)

[84] Bosse, M., Newman, P., Leonard, J., Teller, S.: Slam in large-scale cyclic environments using the atlas framework. International Journal of Robotics Research 23(12), 1113–1139 (2004)

[85] Dissanayake, M., Newman, P., Clark, S., Durrant-Whyte, H., Csorba, M.: A solution to the simultaneous localization and map building (slam) problem. IEEE Transactions on Robotics and Automation 17(3), 229–241 (2001)

[86] Estrada, C., Neira, J., Tardos, J.D.: Hierarchical SLAM: Real-time accurate mapping of large environments. IEEE Transactions on Robotics 21(4), 588–596 (2005)

[87] Guivant, J.E., Nebot, E.M.: Optimization of the simultaneous localization and map-building algorithm for real-time implementation. IEEE Transactions on Robotics and Automation 17(3) (June 2001)

[88] Andrade-Cetto, J., Sanfeliu, A.: Concurrent map building and localization on indoor dynamic environment. International Journal of Pattern Recognition and Artificial Intelligence 16(3), 361–374 (2002)

[89] Liu, Y., Thrun, S.: Results for outdoor-SLAM using sparse extended information filters. In: Proc. IEEE Conference on Robotics and Automation (ICRA), Taipei, pp. 1227–1233 (September 2003)

[90] Davison, A.J., Murray, D.: Simultaneous localization and map-building using active vision. IEEE Transactions on Pattern Analysis and Machine Intelligence 24(7), 865–880 (2002)

[91] Newman, P., Bosse, M., Leonard, J.: Autonomous feature-based exploration. In: Proc. International Conference on Robotics and Automation (ICRA), Taipei, vol. 1, pp. 1234–1240 (September 2003)

[92] Sim, R., Elinas, P., Griffin, M., Little, J.J.: Vision based SLAM using the Rao-Blackwellized particle filter. In: Proc. IJCAI Workshop Reasoning with Uncertainty in Robotics, Edinburgh, Scotland (July 2005)

[93]   Montemerlo, M., Thrun, S., Koller, D., Wegbreit, B.: FastSLAM 2.0: An improved particle filtering algorithm for simultaneous localization and mapping that provably converges. In: Proc. 18th International Joint Conference on Artificial Intelligence (IJCAI), Acapulco, Mexico, pp. 1151–1156 (August 2003)

[94]   Lorigo, L.M., Brooks, A., Grimson, W.E.L.: Visually-guided obstacle avoidance in unstructured environments. In: Proc. IEEE Conference on Intelligent Robots and Systems, France (1997)

[95]   Ulrich, I., Nourbakhsh, I.: Appearance-based obstacle detection with monocular colour vision. In: Proc. AAAI Conference on Artificial Intelligence, USA (July 2000)

[96]   Lenser, S., Veloso, M.: Visual Sonar: Fast obstacle avoidance using monocular vision. In: Proc. IEEE/RSJ International Conference on Intelligent Robots and Systems, Las Vegas, pp. 886–891 (2003)

[97]   Kim, P.G., Park, C.G., Jong, Y.H., Yun, J.H., Mo, E.J., Kim, C.S., Jie, M.S., Hwang, S.C., Lee, K.W.: Obstacle Avoidance of a Mobile Robot Using Vision System and Ultrasonic Sensor. In: Huang, D.-S., Heutte, L., Loog, M. (eds.) ICIC 2007. LNCS, vol. 4681, pp. 545–553. Springer, Heidelberg (2007)

[98]   Bertozzi, M., Broggi, A., Fascioli, A.: Real-time obstacle detection using stereo vision. In: Proc. VIII European Signal Processing Conference, Italy, pp. 1463–1466 (September 1996)

[99]   Badal, S., Ravela, S., Draper, B., Hanson, A.: A practical obstacle detection and avoidance system. In: Proc. 2nd IEEE Workshop on Application of Computer Vision, pp. 97–104 (1994)

[100]   Nirmal Singh, N.: Vision Based Autonomous Navigation of Mobile Robots. Ph.D. Thesis, Jadavpur University, Kolkata, India (2010)

Chapter 2
Interfacing External Peripherals with a Mobile Robot[*]

Abstract. This chapter discusses how real-life interfacing of external peripherals with a ready-made mobile robot can be successfully achieved. Such a system is hoped to be useful for those research scenarios where, many-a-time, because of the fund constraints, a complete robot system cannot be procured with all its accessories and sensor systems. This chapter discusses how such interfacing can be achieved for the KOALA robot using serial communication in interrupt driven mode.

2.1 Introduction

Real-life mobile robots, nowadays, come equipped with several sensors and other accessories which add sophistication and flexibility and help in developing overall capability and intelligence of the system. On many occasions, incorporation of more degrees of automation requires interfacing add-on peripheral devices, which are required to be driven in real life. The robot issues commands for these sensors and accessories time-to-time and these peripheral devices are required to serve the robot's requests, conforming to the demands of a real-time system. Hence a successful development of an integrated system, utilizing the robot core with the add-on components, requires the development of sophisticated interrupt-driven software routines. At present several robotic platforms are available at the disposal of the researchers of the robotic community all over the world, with different degrees of automation. Almost all of them are equipped with several sensors and other accessories with the necessary software support for their intra/inter-communication in real-time. However, these robotic platforms are not that user-friendly, if the user wishes to connect add-on sensors or other accessories as peripheral devices, those are not supported/marketed by the same manufacturing company. Usually the technical knowhow of interfacing such external devices with the mobile robots are also not available, as these details are not provided by the manufacturers, even for those add-on sensors which come along with their robot packages.

---

[*] This chapter is based on: "A PIC Microcontroller system for Real-Life Interfacing of External Peripherals with a Mobile Robot," by N. Nirmal Singh, Amitava Chatterjee, and Anjan Rakshit, published in International Journal of Electronics, vol. 97, issue 2, pp. 139-161, 2010. Reprinted by permission of the publisher (Taylor & Francis Ltd, http://www.tandf.co.uk/journals).

A. Chatterjee et al.: Vision Based Autonomous Robot Navigation, SCI 455, pp. 21–46.
springerlink.com   © Springer-Verlag Berlin Heidelberg 2013

In recent times, PIC microcontroller based systems have found popular real-life applications in several research domains e.g. hardware implementation of the artificial neural network (ANN) model of varicap diode [1], Petri-net based distributed systems for process and condition monitoring [2], development of a double beam modulation system popularly employed in atomic collision experiments [3], hardware implementation of a recurrent neural network model [4], reactive power control of a fuzzy controlled synchronous motor [5] etc. The architecture of PIC microcontrollers is based on a modified Harvard RISC instruction set [6], [1]. These are getting popular day-by-day as they can provide excellent low-cost solutions with state-of-the-art performance. They can provide satisfactory performance because the transfer of data and instruction takes place on separate buses. These processors are also capable of providing increased software code efficiency and simultaneous execution of current instruction with fetching of the next instruction [1]. In this chapter, we shall discuss in detail the development of a PIC microcontroller based system to interface external peripherals with a popular mobile robot available for the research community in the market [10], [11]. The mobile robot under consideration will be KOALA robot from K-team S.A., Switzerland. The KOALA robot procured in our Electrical Measurement and Instrumentation laboratory of the Electrical Engineering Department, Jadavpur University, Kolkata, India, was equipped with incremental encorders, ultrasonic sensors and IR sensors only. However, as our main objective is to use vision sensing for navigation of the KOALA mobile robot, the need was felt to develop and externally integrate and interface a vision sensing system with the robot. Hence the initial research effort was directed to add both stereo-vision facility (comprising two cameras) and mono-vision facility (comprising a single camera) separately with the KOALA robot. This equips the robot with the flexibility of incorporating a two-camera based system or a mono-camera based system for navigation. In this chapter we discuss the addition of stereo-vision facility where four degrees of freedom (DOFs) are added for a vision-system, integrated from outside, with the KOALA robot. The PIC microcontroller based system is developed for pan-control, tilt-control, left-vergence control and right-vergence control of the robot system. The software, developed in interrupt driven mode, is described in detail, which should help other users to develop similar integrated systems. This concept can help to keep complete flexibility at the researcher's/developer's disposal and, at the same time, cost incurred can get drastically reduced. In fact, the main motivation of this research effort was that we could not afford to buy the KOALA robot package with complete integrated vision system due to budgetary constraint and hence the integration of vision system with the KOALA robot was performed by ourselves, in our laboratory. It is sincerely hoped that this effort should encourage other researchers within the robotics community to develop such interrupt-driven systems themselves, which they can utilize to interface stand-alone peripheral devices with other robotic packages as well. This should help in developing low-cost robotic platform with high degrees of sophistication and, although the present system interfaces four peripheral devices (namely four RC servo motors for the four DOFs), the logic can be extended for many more such peripheral devices.

2.2 PIC Microcontroller Based System for Interfacing a Vision System with a Ready-Made Robot

The proposed system employs a PIC 16F876A microcontroller for interfacing the KOALA robot, in real time, with a vision system. The objective here is to interface four RC servomotors [7] with the KOALA mobile robot in real life which can add four degrees of freedom (DOFs) to the vision system, integrated in-house with the KOALA robot. Figure 2.1 shows the complete system where the vision system is integrated with the KOALA robot in our laboratory. It contains the basic KOALA robot with in-built sensors, like infra red sensors and incremental encoder. To increase the capability of the robot system, two ultrasonic sensors and two cameras are additionally integrated to provide the capability of stereo vision. However, as mentioned earlier, the vision-based navigation system developed using KOALA robot is also separately equipped with the capability of mono-camera vision. There is only one significant difference between the system developed using stereo-vision and the system developed using mono-vision. In the case of mono-vision, there is only one camera placed at the center of the active head system and the system utilizes only two DOFs, for pan-control and tilt control. The integrated vision system is so developed that it has the flexibility of controlling four DOFs for stereo vision and two DOFs for mono-vision. Hence, to add extensive flexibility to the vision system, a pan-tilt system is integrated with four servomotors. Figure 2.2 shows the schematic diagram of PIC microcontroller board used to drive four servomotors [7].

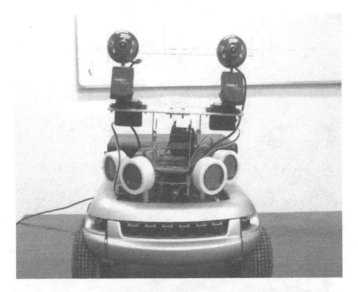

Fig. 2.1. Complete vision system with KOALA robot

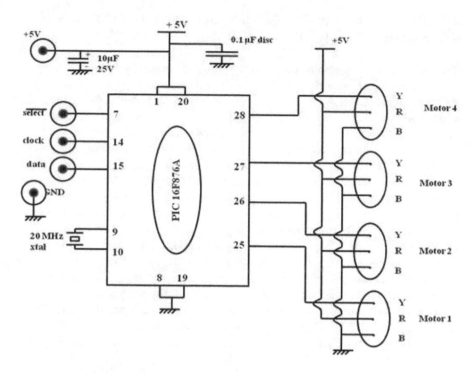

Fig. 2.2. Schematic diagram of the PIC 16F876A based board for interfacing KOALA robot with four servomotors (Y: Yellow; R: Red; B: Black)

Fig. 2.3. Actual photograph of the PIC microcontroller board employed

These four RC servomotors are employed for pan and tilt control of the complete vision system and individual vergence control for each of left camera and right camera. The PIC 16F876A microcontroller receives signal from the Motorola 68331 processor at three input pins, select, clock and data. The 68331 works in SPI master mode and the PIC 16F876A works in SPI slave mode. Figure 2.3 shows the actual photograph of the PIC microcontroller board employed. It is placed in a vertical position against the support of the pan-tilt system, to make the integrated system rugged enough.

As mentioned earlier, this system is employed with PIC 16F876A microcontroller, which is a 28 pin plastic dual-in-line package (PDIP). The key features of the PIC 16F876A microcontroller include [6] 8k flash program memory, 368 bytes data memory and 256 bytes of EEPROM data memory. The operating frequency can vary from DC to 20 MHz and there are provisions for 14 interrupts. PIC 16F876A contains three I/O ports (namely A, B, C), three timers, two analog comparators and five input channels for 10-bit A/D mode. The serial

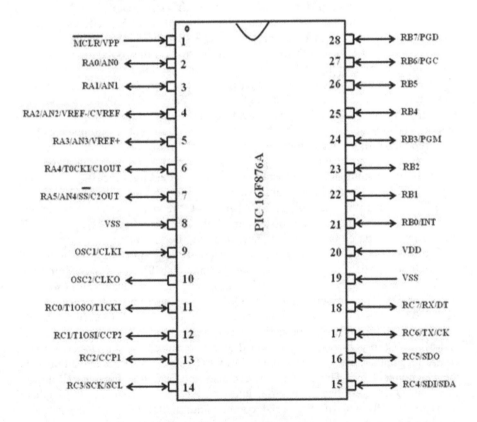

Fig. 2.4. Pin diagram of the 28-pin PDIP PIC 16F876A microcontroller [6]

communications can take place using MSSP and USART. Figure 2.4 shows the pin diagram of the 28-pin PDIP PIC 16F876A microcontroller.

Algorithm 2.1 describes the algorithm for the main PIC microcontroller based program for interfacing external peripherals for real life operation (RC servomotors of the vision system, in this case) with the KOALA robot. This algo. 2.1 describes how the data direction register of PORTB, A/D control register, MSSP control register, INTCON register, PIE1 register and PIR1 registers should be programmed and in which sequence. The system is always initialized so that each RC servo motor is kept initialized at its neutral position and it waits for a suitable input drive command. The system then enables synchronous serial port interrupt and for the PIC microcontroller, the SPI is set in slave mode. Then, depending on interrupt-service-flag content, the corresponding RC servo motor is driven for the specified command. Then the interrupt-service-flag is reset so that it can be made set next time a new interrupt request is placed. Key technical features of the KOALA mobile robot include [8] Motorola 68331 processor with an operating frequency of 22 MHz. The RAM capability of the robot is one Mbyte and the flash capability is one Mbyte. The robot is not equipped with any ROM. KOALA robot is equipped with DC motors with incremental encoders for its motion. A DC motor coupled with the wheel through a 58.5:1 reduction gear is responsible for movement of every wheel. The main processor of the KOALA robot is equipped with the facility of direct control on the motor power supply. The pulses of the incremental encoder can be read by this processor. The RS232 serial link communication is always set at 8 bit, 1 start bit, 2 stop bits and no parity mode. Baud rate can be changed from 9600 baud to 115200 baud. The robot is equipped with 12 digital inputs, 4 CMOS/TTL digital outputs, 8 power digital outputs and 6 analog inputs. The basic module of the robot is equipped with 16 infra-red (IR) proximity and light sensors. These IR sensors embed an IR LED and a receiver. They are manufactured by Texas Instruments (type TSL252) and they can be used for ambient light measurements and reflected light measurements. The output is obtained as an analog value which is converted by a 10 bit ADC. Hence, with this basic arrangement of the KOALA robot, the vision system is integrated, utilizing two cameras, a pan-tilt system, four RC servo motors and the PIC microcontroller based board, used to build the modified robot.

The Motorola MC68331 is a 32-bit microcontroller, equipped with high data-manipulation capabilities with external peripherals [9]. This microcontroller contains a 32-bit CPU, a system integration module, a general-purpose timer (GPT) and a queued serial module (QSM). An important advantage of this MC68331 unit is that it has low power consumption. The CPU is based on the industry-standard MC68000 processor, incorporating many features of MC68010 and MC68020 processors, with so added unique capabilities of high-performance controller applications. A moderate level of CPU control can be achieved utilizing the 11-channel GPT. These GPT pins can also be configured for general-purpose I/O. The QSM comprises two serial interfaces: (i) the queued serial peripheral interface (QSPI) and (ii) the serial communication interface (SCI). The QSPI provides easy peripheral expansion or interprocessor communication.

1.   Initialize PORT B.
2.   Set data direction register corresponding to PORTB such that pin 4-pin 7 of PORTB are configured as digital output.
3.   Set the content of A/D control register 1 such that all pins of the A/D port are configured as digital I/Os.
4.   Set angular position commands to keep each RC servo motor in neutral position i.e. 0^0 position.
5.   Set the content of Master Synchronous Serial Port (MSSP) control register 1 (in SPI mode) such that (i) synchronous serial port is enabled and SCK, SDO, SDI, and $\overline{SS}$ are configured as serial port pins and (ii) SPI is set in Slave mode with $\overline{SS}$ pin control enabled.
6.   Set the content of MSSP status register (in SPI mode) such that the SPI clock select bit is set.
7.   Set the content of peripheral interrupt enable register 1 such that synchronous serial port (SSP) interrupt is enabled.
8.   Set the content of INTCON register to enable all unmasked peripheral interrupts and globally all unmasked interrupts.
9.   Reset the SSP interrupt flag bit of PIR1 register to signify that no SSP interrupt condition has occurred initially.
10.   **IF** interrupt_service_flag is set,

    THEN

        Check the RC servo motor ID for which interrupt occurred.
        Combine contents of 8-bit registers, data-hi and data-lo, to prepare 16-bit angular position command for that servo motor.
        Reset the interrupt_service_flag.

    ENDIF

11.   Set j=1.
12.   **FOR** j=1 to 4,

        Make the corresponding RBx pin of PORTB high, which provides command to RC servo motor (j).
        Keep this RBx pin high for a duration of time, calculated as a function of its corresponding angular position command.
        Then set this RBx pin low.

    ENDFOR
13.   Keep each of RB4-RB7 pins low, for a duration of 20 ms.
14.   Clear CPU watchdog.
15.   Go to step 10.

Algo. 2.1. Algorithm for the main PIC microcontroller based program for interfacing external peripherals, in real life

The algo. 2.1 runs in conjunction with the algo. 2.2, an algorithm developed for synchronous serial interrupt routine. This is a complex and highly sophisticated procedure followed for external peripheral applications in real life. Here the algorithm states how the input drive command is read to perform the specified drive command. The angular position command is composed of two bytes and the high-byte and the low-byte information are transmitted in serial fashion. In fact the first information sent is the RC servo motor id which is required to be driven and this is followed serially by the high-byte and the low-byte of the position command, specifying by how much the RC servo motor should rotate. Figure 2.2 shows that PIC 16F876A uses an external clock of 20MHz frequency. This is the maximum allowable operating frequency for PIC 16F876A processor and the highest permissible value has been utilized to achieve satisfactory performance in this specific application. The PIC processor utilizes three input signals, $\overline{\text{select}}$ (for triggering), clock (for clock signal) and data (for input data waveform to determine a motor drive and its corresponding angular position drive). The PIC processor output is taken from the four pins of the PORTB, which individually produce angular position commands for each RC servomotor.

---

1. **IF** count_flag=0,
    THEN

    > Read the content of the SSP receive/transmit buffer register as the RC servo motor id.

    ENDIF
2. **IF** count_flag=1,
    THEN

    > Read the content of the SSP receive/transmit buffer register as 'data hi', the high byte of the angular position command.

    ENDIF
3. **IF** count_flag=2,
    THEN

    > Read the content of the SSP receive/transmit buffer register as 'data lo', the low byte of the angular position command.

    ENDIF
4. Increment count_flag by 1.
5. **IF** count_flag >2,
    THEN

    > Reset count_flag as 0.
    > Set the interrupt_service_flag.

    ENDIF
6. Reset the SSP interrupt flag bit of PIR1 register.

Algo. 2.2. Algorithm for synchronous serial interrupt service routine

In algo. 2.1, steps 1-9 are employed, for initialization purpose. In the beginning the PORTB is initialized, which is a bidirectional port, of 8 bit width [6].The corresponding data direction register for PORTB is called TRISB. Now, by properly setting the bits of the TRISB register, one can make each corresponding Port B pin, a pin for digital input or digital output. In this work, PORTB is so programmed that its higher four pins, i.e. pin 4 – pin 7 (called RB4-RB7), are configured as digital output. These four pins are used to produce angular position commands for the four servomotors.

Another important module for PIC microcontrollers is the Analog-to-Digital (A/D) converter module. For PIC 16F876A, this module has 5 inputs. Here, this module produces a 10-bit digital number as output for a given analog input signal. This A/D module has four registers i) A/D result high register (ADRESH), ii) A/D result low register (ADRESL), iii) A/D control register 0 (ADCON0) and iv) A/D control register 1 (ADCON1). The ADCON1 register can be used to configure the port pins as analog inputs or as digital I/O. Figure 2.5 shows the description of the ADCON1 register [6]. For the system developed, A/D port configuration control bits are so programmed that these port pins are all configured as digital I/O.

In PIC 16F876A, the serial communication with other peripheral or microcontroller devices is handled by the Master Synchronous Serial Port (MSSP) module. This MSSP is a serial interface which can communicate with serial EEPROMs, ADCs, shift registers etc. The MSSP can operate either in the Serial Peripheral Interface (SPI) mode or Inter-integrated Circuit (I^2C) mode. In this work, the system is developed in SPI mode, where 8 bits of data can be transmitted and received simultaneously in a synchronous manner. The MSSP module uses a status register (SSPSTAT) and two control registers (SSPCON and SSPCON2), in SPI mode of operation. There are two other registers in SPI mode of operation, namely, serial receive/transmit buffer register (SSPBUF) and MSSP shift register (SSPSR). Among these registers, SSPSR is not directly accessible and can only be accessed by addressing the SSPBUF register. In this system, the PIC 16F876A processor works in slave mode and the Motorola 68331 processor in KOALA robot acts in master mode. Figure 2.6(a) and fig. 2.6(b) show the details of the SSPSTAT and SSPCON register. Typically, three pins are utilized for communication: *i*) Serial Data Out (SDO), *ii*) Serial Data In (SDI) and *iii*) serial Clock (SCK). When SPI slave mode of operation is active, a fourth pin, slave select ($\overline{SS}$) is also used. In the SPI mode of operation, SSPCON register is so programmed that the SSPEN bit is made high, which enables the serial port and configures SCK, SDO, SDI and ($\overline{SS}$) pins as serial port pins. Similarly SSPM3:SSPM0 bits are so programmed that the PIC processor is configured in SPI slave mode, with its clock being assigned to the SCK pin. Hence, in Fig. 2.2, select corresponds to the $\overline{SS}$ pin, clock signal arrives at SCK pin and data arrives at SDI pin. The SSPSTAT register is so programmed that the CKE bit is made high. This ensures that the transmission of data occurs on transition from active to idle clock state. Figure 2.7 shows the SPI Master/Slave connection programmed in this system to interface the Motorola processor with the PIC processor. As the PIC processor is programmed in slave mode, in receive operations, SSPSR and SSPBUF together create a double-buffered receiver. Here SSPSR shift register is used to shift data in or out (MSB first) and SSPBUF is the buffer register in which data bytes are either written or data bytes are read from it.

ADFM	ADCS2	-	-	PCFG3	PCFG2	PCFG1	PCFG0

bit 7   bit 0

bit 7   : ADFM (A/D Result Format Select Bit)
bit 6   : ADCS2 (A/D Conversion Clock Select Bit)
bit 5-4 : Unimplemented
bit 3-0 : PCFG3-PCFG0 (A/D Port Configuration Control Bit)

Fig. 2.5. The description of the ADCON1 register [6]

SMP	CKE	D/$\overline{\text{A}}$	P	S	R/$\overline{\text{W}}$	UA	BF

bit 7   bit 0

bit 7: SMP (Sample bit)
bit 6: CKE (SPI Clock select bit)
bit 5: D/$\overline{\text{A}}$ (Data/$\overline{\text{Address}}$ bit)
bit 4: P (Stop bit)
bit 3: S (Start bit)
bit 2: R/$\overline{\text{W}}$ (Read/$\overline{\text{Write}}$ bit information)
bit 1: UA (Update Address bit)
bit 0: BF (Buffer Full status bit)
(Receive mode only)

Fig. 2.6 (a).

WCOL	SSPOV	SSPEN	CKP	SSPM3	SSPM2	SSPM1	SSPM0

bit 7   bit 0

bit 7   : WCOL (Write collision detect bit)
            (Transmit mode only)
bit 6   : SSPOV (Receive overflow indicator bit)
bit 5   : SSPEN (Synchronous serial port enable bit)
bit 4   : CKP (clock polarity select bit)
bit 3   : SSPM3-SSPM0 (Synchronous serial port mode select bits)

Fig. 2.6(b).

Fig. 2.6(a) & (b). The details of the SSPSTAT and SSPCON register (SPI mode) [6]

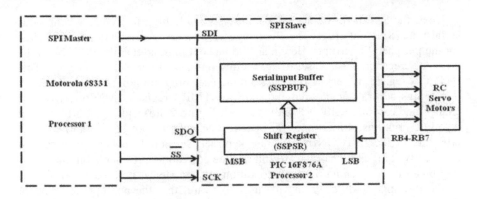

Fig. 2.7. SPI Master/Slave connection programmed in our system

PSPIE	ADIE	RCIE	TXIE	SSPIE	CCP1IE	TMR2IE	TMR1IE

bit 7   bit 0

bit 7 : PSPIE (Parallel slave port read/write interrupt enable bit)
bit 6 : ADIE (A/D converter interrupt enable bit)
bit 5 : RCIE (USART Receive interrupt enable bit)
bit 4 : TXIE (USART Transmit interrupt enable bit)
bit 3 : SSPIE (SSP interrupt enable bit)
bit 2 : CCP1IE (Interrupt enable bit)
bit 1 : TMR2IE (TMR2 to PR2 match interrupt enable bit)
bit 0 : TMR1IE (TMR1 overflow interrupt enable bit)

Fig. 2.8(a)

GIE	PEIE	TMR0IE	INTE	RBIE	TMR0IF	INTF	RBIF

bit 7   bit 0

bit 7 : GIE (Global interrupt enable bit)
bit 6 : PEIE (Peripheral interrupt enable bit)
bit 5 : TMR0IE (TMR0 overflow interrupt enable bit)
bit 4 : INTE (RB0/INT External interrupt enable bit)
bit 3 : RBIE (RB Port change interrupt enable bit)
bit 2 : TMR0IF (TMR0 overflow interrupt flag bit)
bit 1 : INTF (RB0/INT External interrupt flag bit)
bit 0 : RBIF (RB Port change interrupt flag bit)

Fig. 2.8(b)

Fig. 2.8(a) & (b). Details of PIE1 and INTCON register respectively

Next, the SSPIE bit of the PIE1 register is made high to enable synchronous serial port (SSP) interrupt. The PIE1 register contains the individual enable bits for the peripheral interrupts. Here it should be kept in mind that the PEIE bit of the INTCON register must be set to enable any unmasked peripheral interrupt. This INTCON register is very important from the user program point of view, because it contains several enable and flag bits for TMR0 register overflow, RB port change and external RB0/INT pin interrupts. Figure 2.8(a) and Fig. 2.8(b) present detail descriptions of PIE1 and INTCON registers. In the program, both PEIE and GIE bits of the INTCON register are kept set. It should be kept in mind that irrespective of the states of the GIE bit and the corresponding enable bit, interrupts flag bits are set when an interrupt condition occurs. Hence it is the responsibility of the developer, while writing the user software, that the appropriate interrupt flag bits must be reset before an interrupt is enabled. Hence initially the SSPIF bit of the PIR1 register is reset to signify that no SSP interrupt condition has occurred and the system is kept ready to enable the SSP interrupt. The PIR1 register is a special register that contains the individual flag bits for the peripheral interrupts. Hence, the user software must be so written that, before enabling a specific interrupt, the corresponding flag bit in the PIR1 register must be reset. Figure 2.9 shows the details of the PIR1 register.

Once the initialization phase is completed, steps 10-15 in the main program, described in algo. 2.1, execute an infinite loop, in conjunction with the serial interrupt service routine, given in algo. 2.2. Within the interrupt routine, the PIC processor waits for data written/transmitted by the Motorola processor in SPI master mode. Figure 2.10 shows the form of SPI write received. The waveforms show data write of one byte and this process is repeated for each byte written i.e. each byte received by the PIC processor. The system is so programmed that each data bit can be latched either on rising edge or falling edge of the clock signal. The data transfer is always initiated by the Motorola processor in master mode, by sending the SCK signal. When the $\overline{SS}$ pin of the PIC processor is low, then transmit and receive operations are enabled. Then, in the slave mode, the SPI module will be reset if the $\overline{SS}$ pin is set high, or by clearing the SSPEN bit. In our system, this SPI module is reset by forcing $\overline{SS}$ pin to high. The SSPBUF holds the data that was written to the SSPSR until the received data is ready. As mentioned

PSPIF	ADIF	RCIF	TXIF	SSPIF	CCP1IF	TMR2IF	TMR1IF
bit 7							bit 0

bit 7 : PSPIF (Parallel slave port read/write interrupt flag bit)
bit 6 : ADIF (A/D Converter interrupt flag bit)
bit 5 : RCIF (USART Receive interrupt flag bit)
bit 4 : TXIF (USART Transmit interrupt flag bit)
bit 3 : SSPIF (Synchronous serial port (SSP) interrupt flag bit)
bit 2 : CCP1IF (CCP1 interrupt flag bit)
bit 1 : TMR2IF (TMR2 to PR2 match interrupt flag bit)
bit 0 : TMR1IF (TMR1 overflow interrupt flag bit)

Fig. 2.9. Details of the PIR1 registe

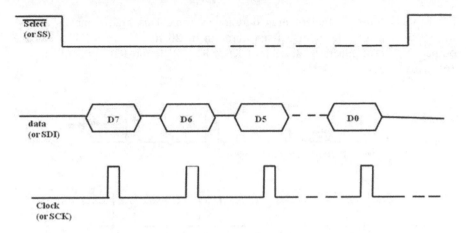

Fig. 2.10. SPI Mode waveform (slave mode)

earlier, data is received byte by byte, and, for each byte received, once the eight bits of the data have been received, that byte is moved to the SSPBUF register. This operation is marked by making the buffer full detect bit (BF) and the interrupt flag bit (SSPIF) high. Hence this double buffering scheme enables to start receiving new data byte before completely reading the data byte that was just received. The sequence of data bytes transmitted by the Motorola processor and hence received, in this serial interrupt mode, by the PIC processor, is programmed as:

a)   Send data byte with RC servomotor ID (between 1 and 4).
b)   Allow a delay of 10 ms.
c)   Send the high data byte corresponding to the angular position command for that specific RC servomotor ID.
d)   Allow a delay of 10 ms.
e)   Send the low data byte corresponding to the angular position command for the same RC servo motor.

In each of steps (a), (c) and (e), the content of SSPBUF register is read in different temporary variables. When a sequence of (a) to (e) is completed, a complete information transmission cycle takes place. This is marked by setting a temporary flag variable (interrupt_service_flag) in the software. The user program must also reset the SSP interrupt flag bit (SSPIF) of the PIR1 register before returning from the Interrupt Service Routine. Once this interrupt_service_flag is set in the main program, the high byte and the low byte of the RC servo motor position command are combined to create an appropriate position command and a corresponding drive command is sent to the RC servo motor. For each RC servo motor, this digital drive is given by driving the corresponding pin signal high for a certain period and then forcing the same pin signal low for a certain period. The high signal period of 1.5 ms corresponds to an angular command of 0^0, 1 ms corresponds to -90^0 and 2 ms to $+90^0$. For all intermediate angle commands, a proportional timing signal of high duration is sent. This high signal is always followed by a low signal of a fixed duration of 20 ms. Figure 2.11 shows the output waveform generated at each of RB4, RB5, RB6 and RB7 pins of the PIC 16F876A processor.

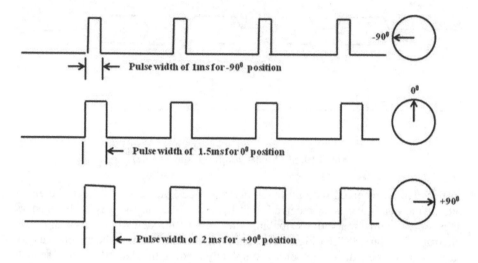

Fig. 2.11. Output waveform generated at RB4, RB5, RB6 and RB7 pins of the PIC 16F876A processor

As mentioned, in this implementation the SPI mode is utilized and not I^2C mode for serial communication. It is well known that I^2C mode is more flexible with the facility of multi-drop bus type architecture but the system becomes more complicated with higher degree of sophistication involved. On the other hand, SPI mode is suited for only single drop, point-to-point architecture and the system is less complicated. The SPI mode is more suitable for the system developed here as a point-to-point communication with the KOALA robot is needed only and it helps to keep the system less complicated, specially for real-life communication.

2.3 The Integrated System Employing KOALA Robot with a PC and a Vision System

Figure 2.12 shows the complete integrated system with a PC-based KOALA robot that communicates with the RC servo motors of the vision system, through the PIC 16F876A microcontroller based board. The PIC microcontroller communicates with four external peripherals, those are four servomotors of the vision system. These four servomotors are employed for pan control, tilt control, vergence control of the left camera and vergence control of the right camera of the vision system. The addition of these four degrees of freedom (DOFs) to the vision system and their efficient control adds flexibility and high degree of automation to the entire integrated system.

A PC based system is also developed for communication with the KOALA mobile robot. This is a GUI based system developed that communicates with the robot by sending ASCII strings of commands and it can also accept sensor readings returned from the robot. Figure 2.13 shows a snapshot of the form developed that interacts with the user. The user is provided with the provision of keying in the driving command which is sent in serial mode of transmission. The GUI based system transmits the ASCII string typed, in serial mode, when the 'Send' button is clicked. The system can also display whatever data is received from the robot in a different display box. This display box can help us to check whether the system under control is performing the commanded task in a desired manner.

For the KOALA robot end, another C program is developed and its cross-compiled version (with .s37 extension) is downloaded in the Motorola processor. This .s37 program also communicates with the VB program in the PC end in the interrupt mode where it always expects a driving command, sent in form of an ASCII message and serves this driving command in a highly sophisticated manner. For this serial communication mode, the host PC plays the role of the master and the KOALA robot plays the role of the slave. Every interaction between the host PC and the KOALA, which is configured as a remote terminal unit (RTU), takes place in the following manner [8]:

- A user defined ASCII string, terminated by a carriage return (CR), is sent by the host PC to the robot.
- If the host PC command the robot to acquire and send some sensor readings (e.g. IR sensors, ultrasonic sensors etc), the KOALA robot responds by sending back the sensor readings in form of an ASCII message, terminated by a carriage return (CR).

Figure 2.14 shows a simple example of a .c example program written for the KOALA robot to turn around. Figure 2.15 shows the sequence to be followed so that this program, developed in PC, can be cross compiled to generate an .s37 file and can be downloaded in the Motorola processor of the KOALA robot.

The KOALA robot, in its basic package, already supports a pool of tools and protocol commands, where one can open a terminal emulator in the host PC, with the serial communication protocol set, and one can execute the protocol commands. These protocol commands include very useful commands like set speed ('D'), read speed ('E'), read A/D input ('I'), read management sensors ('M'), read proximity sensors ('N') etc. However, in addition to these KOALA supported commands, other protocol commands are also created to communicate with peripherals integrated additionally. Hence, a whole pool of protocol commands is created, to bring uniformity in the way the robot is going to be commanded from the host PC. These commands include some commands which

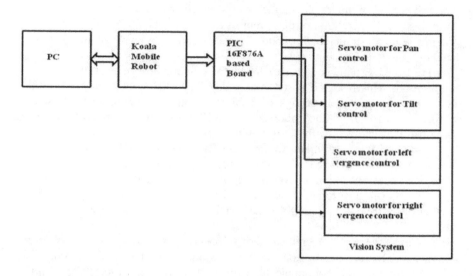

Fig. 2.12. The integrated system employing KOALA Robot with a PC and a vision system

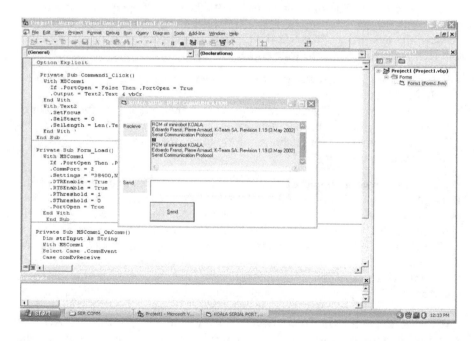

Fig. 2.13. Snapshot of the form developed that interacts with the user

```
#include <sys/Kos.h>
int
main ()
{
        mot_reset();
        mot_config_speed_1m (0, 1000, 800, 100);
        mot_config_speed_1m (1, 1000, 800, 100);
        mot_new_speed_1m (0, -6);
        mot_new_speed_1m (1, 6);
        return 0;

}
```

Fig. 2.14. An example .c program for turning the KOALA robot around

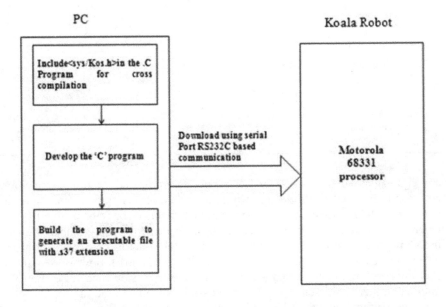

Fig. 2.15. A block diagram representation of the sequence of operations followed to download the cross compiled version of a .c program

Table 2.1. Command protocol for serial communication between host PC and KOALA Robot

Nature of command	Command Protocol (ASCII Message string)	Command class
Set KOALA Motor Speed	\<D>\<s>\<ddd>\<s>\<ddd>\<CR>	KOALA Equivalent command
Set KOALA position reached	\<C>\<s>\<ddddd>\<s>\<ddddd>\<CR>	- do -
Request IR proximity sensor data	\<N>\<CR>	
Request IR ambient light sensor data	\<O>\<CR>	- do -
Request ultrasonic sensor data	\<I>\<CR>	- do -
Request the speed of the motors	\<E>\<CR>	- do -
Set the position of the motor	\<G>\<CR>	- do -
Request the position of the motor	\<H>\<CR>	- do -
Set RC Servo motor position command	\<Z>\<i>\<s>\<dd>\<CR>	Add-on Peripheral Interface Command

are equivalent versions of KOALA supported commands and the remaining commands are created additionally within the scope of this work, to communicate with the add-on peripherals. Another important point to be kept in mind is that when the KOALA supported commands are executed from the terminal emulator in PC, it goes to the monitor program module in the robot processor and performs its designated function. However, when a .s37 program is downloaded in the Motorola processor to communicate with the PC side and also the PIC microcontroller side, then it will not be possible to activate the monitor program from the terminal emulator. Hence it is very important to bring all robot protocol commands under one roof (i.e. activated from the VB program and executed by the .s37 program). Hence the C program written for the KOALA robot had to include actions for all such ASCII request messages, sent from the PC end. The sample list of commands and their formats of ASCII messages sent by the host are presented in Table 2.1.

In this Table 2.1, e.g., to set KOALA motor speed, one can start with the identifier character 'D', followed by speed of each motor, set as a three digit number (KOALA is a differential drive system). The speed command for 'motor 0' is followed by the speed command of 'motor 1'. As the polarity of speed of each individual motor in the KOALA robot can be set separately for forward motion or backward motion, each speed information for individual motor is preceded by a sign, shown as <s> in the command protocol. This is set as '+' for forward motion of the motor and as '-' for backward motion of the motor. When the Motorola processor receives this ASCII message string, the .s37 program developed performs a suitable decoding of the string and executes the command by driving each motor of the KOALA robot according to the protocol command. Similar actions are performed for each KOALA equivalent command, issued from the PC end, as an ASCII message string.

However, when the Motorola processor receives the ASCII message string corresponding to the add-on Peripheral Interface Command, it suitably decodes the protocol command and sends appropriate drive command to the PIC processor. In this situation, the Motorola processor acts in master mode and the PIC processor acts in slave mode, as shown previously in Fig. 2.7. The message string for this action, sent from the PC end, comprise 'Z' as the identifier character, followed by <i>, which corresponds to the servomotor id for which position command is prepared (i can vary from 1 to 4), followed by the sign, given in <s>, for position command ('+' or '-') and then the two digit actual position command, in degree (this can vary from -90^0 to $+90^0$). Every ASCII message string is terminated, as usual, by using carriage return, <CR>. Once this string is received by the Motorola processor, it decodes the string to produce a byte information for motor id and a word information for the corresponding angular position command (1500 μs for 0^0; 1000 μs for -90^0; 2000 μs for $+90^0$; proportional interpolated timing values for each intermediate angular command). This word information is then decomposed to separately produce corresponding high byte information and low byte information.

2.4 Real-Life Performance Evaluation

The performance evaluation has been carried out for all the protocol commands by experimentally issuing ASCII message commands from the host PC terminal (i.e. from the VB end) to the KOALA robot through serial communication. Before starting the experimental evaluation, the calibration of each RC servo motor is separately individually tested and software corrections are introduced for each of them to take care of small offsets. A series of experiments has been carried out to test that the integrated robotic system actually performs the task commanded by the ASCII message string. Some of these sample experimental cases are described below.

In one sample case, for testing the driving motion, the robot was commanded to move in forward direction with a small uniform speed for both robots. Figure 2.16 shows the snapshot of the GUI-based system at the PC end where we issued the ASCII message string "D+002+002". It was found that the robot moved in conformity with the command issued. Similarly these commands were tested with higher motor speed commands, with negative motion commands for reverse movement and with differential drive commands for each motor. It was found that the integrated robotic system showed ordered behavior in accordance with each command issued.

Another set of case studies was performed, where the two-way serial communication between the host PC and the KOALA robot was tested. In these experiments, the robot was commanded to acquire sensor readings and then return them to the host PC end. The ASCII command 'N' was issued from the PC end to acquire the readings of all the sixteen infrared proximity sensors, placed around the robot. The infrared sensors are used for a range of 5 cm. to 20 cm. The output of each measurement given by an infrared sensor is an analog value, which is digitized by a 10 bit ADC. An obstacle of 5 cm. width was placed in front of the IR sensors L0 and R0 (the two IR sensors which are directly at the axial heading positions of the robot) at two different distances of 5 cm. and 20 cm. respectively. In conformity with the actual situation, the sensor readings returned gave higher values for smaller distances. When the distance was only 5 cm. the reading returned by both L0 and R0 are 1023, the maximum possible value. The readings of the other IR sensors are also obtained in form of a string and displayed at the host PC end, as shown in Fig. 2.17(a). It can be seen that for the other IR sensors, for this given position of the obstacle, the readings obtained were smaller as the distance between each sensor and the obstacle was more than the distance of the obstacle from each of L0 and R0. When the distance between the obstacle and the heading direction of the robot was increased to 20 cm., the readings obtained from L0 and R0 were 253 and 348, as is shown in Fig. 2.17(b), and accordingly, the readings obtained from other sensors were also significantly reduced in values. Similarly, the readings of the two ultrasonic sensors were obtained by issuing ASCII 'I' command from the PC end. These sensors can be used for obstacle detection over a range of 15 cm. to 300 cm. The corresponding analog output of the sonar sensor varies in the range 0 volt to 4.096 volts. A corresponding mapping in the digital form is carried out in the range 0-1023 where the zero corresponds to 0 volts (minimum distance) and 1023 correspond to 4.096 volts (maximum distance). The readings obtained for two sample case studies, where the robot was placed in front of a wall at a distance of 100 cm. and 200cm. respectively, are shown in Fig. 2.18(a) and Fig. 2.18(b) respectively. When the wall was at a distance of 100 cm, the values of the sonar sensors returned were 403 and 417. Similarly, when the wall was at a distance of 200 cm, the values of the sonar sensors returned were much higher, found to be 885 and 895.

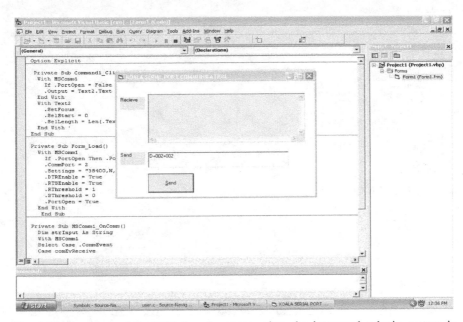

Fig. 2.16. Snapshot of the GUI based system, when the integrated robotic system is commanded for a driving motion in forward direction

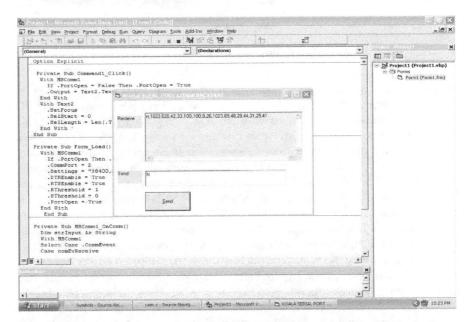

Fig. 2.17(a).

Fig. 2.17(a) & (b). Snapshot of the GUI based system, when the integrated robotic system is commanded to acquire IR sensor values, at a distance of 5 cm. and 20 cm. respectively, from an obstacle of 5 cm. width

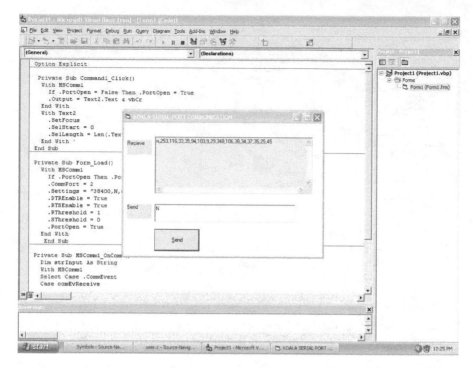

Fig. 2.17(b).

Fig. 2.17(a) & (b). (*continued*)

Another very important set of experimentations carried out was for the add-on peripheral interface command, where the serial command issued from the host PC end (acting in master mode) starts with the ASCII character identifier 'Z'. This command is received by the Motorola processor of the KOALA robot in slave mode and subsequently the Motorola processor (now acting in master mode) commands each servomotor, initiating serial communication through the PIC microcontroller board, where PIC 16F876A processor acts in slave mode. Figure 2.19(a) shows the initial condition, where each servomotor is at its neutral position. Then four commands, 'Z1+45', 'Z2+45', 'Z3+45', and 'Z4+45', were issued, separately, sequentially, so that each servomotor for vergence control of left camera, vergence control of right camera, pan control, and tilt control, is commanded to rotate by an angle of 45°, in a sequence. Fig. 2.19(b)-2.19(e) show the snapshots of the system acquired after issuing each such command from the host PC end. These experimentations show that these add-on peripherals, integrated with the KOALA robot, are successfully interfaced for real life applications and could be suitably commanded from the host PC end, as desired. Table 2.2 shows the time delay in issuing a command from the PC-end (where the user issues the command) and each RC servo-motor performing its function

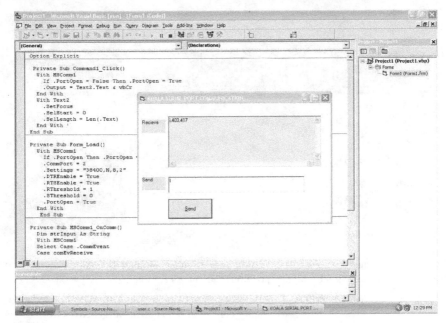

Fig. 2.18(a).

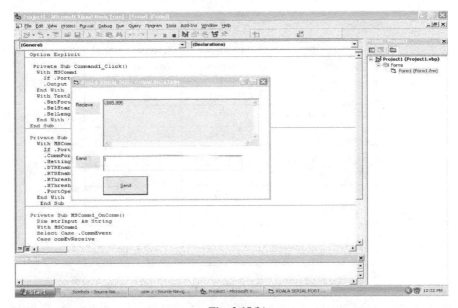

Fig. 2.18(b).

Fig. 2.18(a) & (b). Snapshot of the GUI based system, when the integrated robotic system is commanded to acquire ultrasonic sensor values, at a distance of 100 cm. and 200 cm. respectively, from the wall

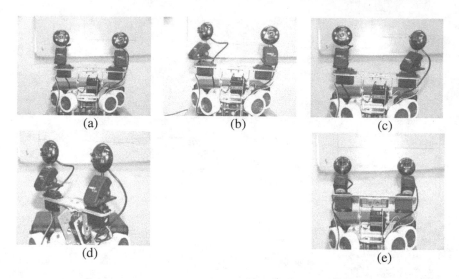

(a)   (b)   (c)
(d)   (e)

Fig. 2.19(a). Snapshot of the system configuration under initialized condition i.e. each of the four servomotors is at 0^0 angular position (called the neutral position)

Fig. 2.19(b)-2.19(e). Snapshots of the system position for angular commands of 45^0 given to each of the RC servomotors for vergence control of left camera, vergence control of right camera, pan control and tilt control, each command issued from the PC end sequentially

Table 2.2. Sensing and reacting delay time for RC servo motors

Degrees of freedom	Angular command issued (degree)	Sensing and Reacting delay time (seconds)
Left vergence axis	90	0.433
	45	0.200
Right vergence axis	90	0.433
	45	0.200
Pan axis	90	0.633
	45	0.333
Tilt axis	90	0.400
	45	0.300

completely as commanded. The experimentations were carried out for each RC servo motor with two angular commands i.e. 45° and 90°. It can be seen that the left vergence, right vergence and tilt axis showed very similar sensing and reacting delays. The pan axis showed a little more delay in real life compared to the other three degrees of freedom.

2.5 Summary

In this chapter, we discussed and demonstrated in detail how a PIC microcontroller based system can be developed for real-life interfacing of external peripherals with a ready-made mobile robot, in this case the KOALA robot. For the system described here, the serial communication is developed in interrupt driven mode, where the KOALA processor acts in master mode and the PIC processor acts in slave mode. A complete integrated system is developed in house in our laboratory employing a PC, the KOALA robot, the PIC microcontroller based board and a two-camera based vision system. Here the PIC microcontroller based system serves four external peripherals, i.e. four RC servomotors, included to control four degrees of freedom of the vision system. The complete system works under the control of a PC-based GUI system, where the system at PC end acts in master mode and performs serial communication, under interrupt driven mode, with the Motorola processor in KOALA robot, acting in slave mode.

The developed system demonstrated how real-life add-on peripherals can be integrated from outside with a basic robotic platform to enhance capability and sophistication of the integrated system, developed at a much lesser cost. The concept presented here can be extended for adding many other external peripherals and the concept should also be useful for other real mobile robots. It is also hoped that a detailed discussion on development of such systems, as presented in this chapter should help other researchers in Robotics community to develop similar integrated systems. This should be quite useful in the research domain, as usually these technical knowhow of integrating sensors and other components within an integrated robotic platform remains with the manufacturer and usually it is not available in public domain. It should also be mentioned that although all experimentations shown here are carried out for two-camera based system, the system is equally applicable for a mono-camera based system, where the camera can be placed at the center of the active head and, in that case, controlling only two DOFs (for pan and tilt only) will be enough.

Acknowledgement. The work described in this chapter was supported by University Grants Commission, India, under Major Research Project Scheme (Grant No. 32-118/2006(SR)).

References

[1] Turkoglu, I.: Hardware implementation of varicap diode's ANN model using PIC microcontrollers. Sensors and Actuators A: Physical 138, 288–293 (2007)

[2] Frankowiak, M.R., Grosvenor, R.I., Prickett, P.W.: A Petri-net based distributed monitoring system using PIC microcontrollers. Microprocessors and Microsystems 29, 189–196 (2005)

[3] O'Neill, R.W., Greenwood, J.B., Gradziel, M.L., Williams, D.: Microcontroller based double beam modulation system for atomic scattering experiments. Measurement Science and Technology 12, 1480–1485 (2001)

[4] Hung, D.L., Wang, J.: Digital hardware realization of a recurrent neural network for solving the assignment problem. Neurocomputing 51, 447–461 (2003)

 [5] Colak, I., Bayindir, R., Sefa, I.: Experimental study on reactive power compensation
       using a fuzzy logic controlled synchronous motor. Energy Conversion Management
       45(15-16), 2371–2391 (2004)
 [6] PIC 16F876A Data sheet, 28/40/44- Pin Enhanced Flash Microcontrollers.
       Microchip Technology Inc. (2003)
 [7] HS-322, H.R.: Servo Motor manual, HITEC RCD KOREA Inc. (2002)
 [8] KOALA User Manual, Version 2.0(silver edition). K-team S.A., Switzerland (2001)
 [9] MC68331 User Manual, MOTOROLA, INC. (1996)
[10] Nirmal Singh, N., Chatterjee, A., Rakshit, A.: A PIC Microcontroller system for
       Real-Life Interfacing of External Peripherals with a Mobile Robot. International
       Journal of Electronics 97(2), 139–161 (2010)
[11] Nirmal Singh, N.: Vision Based Autonomous Navigation of Mobile Robots. Ph.D.
       Thesis, Jadavpur University, Kolkata, India (2010)

Chapter 3
Vision-Based Mobile Robot Navigation Using Subgoals[*]

Abstract. This chapter discusses how a vision based robot navigation scheme can be developed, in a two-layered architecture, in collaboration with IR sensors. The algorithm employs a subgoal based scheme where the attempt is made to follow the shortest path to reach the final goal and also simultaneously achieve the desired obstacle avoidance. The algorithm operates in an iterative fashion with the objective of creating the next subgoal and navigating upto this point in a single iteration such that the final goal is reached in minimum number of iterations, as far as practicable.

3.1 Introduction

Recent advances in technologies in the area of robotics have made enormous contributions in many industrial and social domains. Nowadays numerous applications of robotic systems can be found in factory automation, surveillance systems, quality control systems, AGVs (autonomous guided vehicles), disaster fighting, medical assistance etc. More and more robotic applications are now aimed at improving our day-to-day lives, and robots can be seen more often than ever before performing various tasks in disguise [1]. For many such applications, autonomous mobility of robots is a mandatory key issue. Many modern robotic applications now employ computer vision as the primary sensing mechanism. As mentioned earlier in this book, vision system is considered as a passive sensor and possesses some fundamental advantages over the active sensors such as infrared, laser, and sonar sensors. Passive sensors such as cameras do not alter the environment by emitting lights or waves in the process of acquiring data, and also the obtained image contains more information (i.e. substantial, spatial and temporal information) than active sensors [2]. Vision is the sense that enables humans to extract relevant information about the physical world, and appropriately it is the sense that we, the humans, rely on most. Computer vision

---

[*] This chapter is adopted from Measurement, vol. 44, issue 4, May 2011, N. Nirmal Singh, Avishek Chatterjee, Amitava Chatterjee, and Anjan Rakshit, "A two-layered subgoal based mobile robot navigation algorithm with vision system and IR sensors," pp. 620-641, © 2011, with permission from Elsevier.

techniques capable of extracting such information are continuously being developed and more and more real-time vision-based navigation systems for mobile robots are being implemented now.

Vision based Robot navigation is defined as the technique that guides a mobile robot to a desired destination, or along a desired path in an environment, by avoiding static (and may be dynamic) obstacles primarily using vision sensor [3], [4]. In this chapter, we describe the real-life implementation of a mobile robot navigation scheme, where vision sensing is employed as primary sensor for path planning and IR sensors are employed as secondary sensors for actual navigation of the mobile robot with obstacle avoidance capability in a static or dynamic indoor environment. As described previously, the popular choices for the creation of the environment maps can be grid-based [5, 6, 7], topological map [8, 9], hybrid map [10] etc. The mapless navigation systems are those that use no explicit representation at all of the space in which navigation is to take place and they rather resort to recognizing objects found in the environment or to tracking or avoiding those objects by generating motion commands based on visual observations [11, 12]. Several research works have so far been reported to acquire knowledge about the environment using camera(s) in stereo vision [13, 14], trinocular vision [15], omni-directional or panoramic vision [16, 17], and monocular vision [18, 19]. Each such solution in mobile robot navigation has its own advantages and disadvantages. In those situations where the knowledge of the map is available, an important problem in navigation is the path planning for intelligent control or guidance of the mobile robot. The popular general approaches for path planning can be based on roadmap, cell decomposition, potential field etc. [20]. They differ in how the connectivity graphs are constructed and their representations. Obviously, without any *a priori* knowledge of an environment, it is almost impossible to determine the true shortest path for navigation, among all possible paths. It is potentially possible to determine such paths by employing standard graph-search techniques, such as Dijkstra's algorithm [21] and A* algorithm [22].

As mentioned earlier, in this chapter we describe a goal driven approach for mobile robot navigation, using vision based sensing and IR sensor based navigation [28, 29]. This two-layer based approach attempts to determine the shortest path of navigation between the start point and the known goal point, given a static or dynamic environment, in presence of obstacles. In the first layer, vision acts as the primary sensing system to acquire image of the environment, for subsequent path planning. A series of image processing operations is performed on the acquired image and then a gradient descent based algorithm is employed to compute the shortest path between the present position of the robot and the goal, avoiding obstacles [26]. This shortest path is employed to generate a subgoal and this information is then locally utilized to navigate the robot, utilizing IR sensor based guidance. This second layer of IR sensor based robot navigation attempts to guide the robot to the subgoal, even if the environment changes dynamically. Once the robot reaches the subgoal, the two-layer based algorithm is again activated to generate a new subgoal and to navigate the robot till this new subgoal is reached. This process is repeated iteratively until the final goal is reached. This method simultaneously attempts to attain two objectives. Based on vision sensing, it attempts to implement a shortest path planning algorithm in a bid to reach the goal, avoiding obstacles, as fast as it can. Then, if the environment undergoes a change during navigation and obstacle information gets updated, then IR sensor based guidance equips the robot with the capability of handling the changed environment so that the robot can still navigate safely. The periodic usage of vision based updating of the environment, subsequent path planning and then IR based actual navigation helps to guide the robot to adapt its navigation temporally with dynamic variations in the environment and still attempt to reach the goal in shortest time, as quickly as practicable. This algorithm was implemented in our laboratory, for the KOALA robot [23], creating several real-life like environments. The results showed the usefulness of the proposed algorithm. The algorithm is described in detail in subsequent sections of this chapter.

3.2 The Hardware Setup

The KOALA robot was described in detail in the previous chapter. Still we recapitulate salient features of the KOALA robot to provide a brief introduction of the hardware setup utilized for this real-life implementation carried out. KOALA is a small (32 cm x 32 cm) six wheeled, differential drive vehicle manufactured by K-team, Switzerland [23]. The KOALA robot used in our laboratory is equipped with 16-proximity/ambient IR sensors, four ultrasonic sensors and wheel encoders. We have integrated two complete vision systems along with the KOALA robot in our Electrical Measurement and Instrumentation Laboratory, Electrical Engineering Department, Jadavpur University, Kolkata. The vision system is so developed that it can work either with a stereo vision system employing two cameras (as described in the previous chapter) or it can employ a single camera based system. The algorithm that we describe now is based on employing a single wireless camera for monocular vision. In KOALA, the hardware control is performed by an on- board microprocessor (Motorola 68331@ operating frequency 22MHz) [23]. Figure 3.1(a) shows a snapshot of the mobile robot with four ultrasonic sensors and the vision system, integrated in our laboratory, employing a single vision sensor. The ultrasonic sensors can detect obstacles over a wide range from 15 cm to 300 cm, and the IR sensors will provide a range of measurements from 5 cm to 20 cm. Our system utilizes single vision sensor comprising a JMK wireless camera (WS-309AS) with A/V receiver and a Frontech USB (TV Box) frame grabber, which is used for acquiring a running video stream. Figure 3.1(b) shows the vision system in schematic form. The entire system is developed with an objective of providing a low-cost solution which should prove attractive for the industrial community. This monocular vision system is developed with two degrees of freedom to provide pan control and tilt control. To add two degrees of freedom (DOFs) for this vision-system, a PIC (16F876A) microcontroller based system is developed in our laboratory for pan-control and tilt-control of the single-camera based robot system [24]. Here, the main onboard Motorola microcontroller acts as the master and the PIC microcontroller acts as a slave. The software, developed in interrupt driven mode, communicates with the mobile robot through the RS232C port. Figure 3.2 shows a snapshot of the user-interface developed in the PC side that can interact with the user. The main serial mode of communication is handled by passing ASCII message strings between the PC and the Motorola processor in the robot.

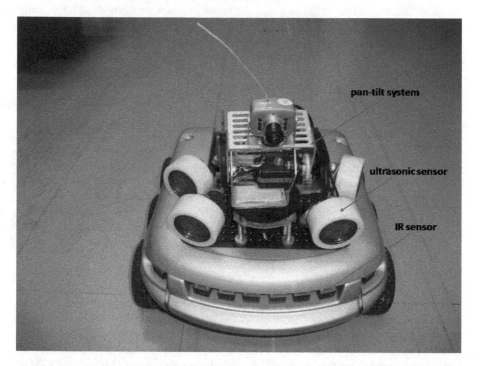

Fig. 3.1(a). The KOALA robot, equipped with sonar and IR sensors and integrated with a single camera based vision system

The RS 232C serial link set-up between the PC and the robot is always set at 8 bit data, 1 start bit, 2 stop bits and no parity mode. To give an example, the message string for RC servo action to provide pan or tilt control, sent from the PC end, comprise 'Z' as the identifier character, followed by $<i>$, which corresponds to the servomotor id for which position command is prepared ($i = 1$, 2), followed by the sign, given in $<s>$, for position command ('+' or '-') and then the two digit actual position command, in degree (this can vary from -90^0 to $+90^0$). Every ASCII message string is terminated, as usual, by using carriage return, $<CR>$.

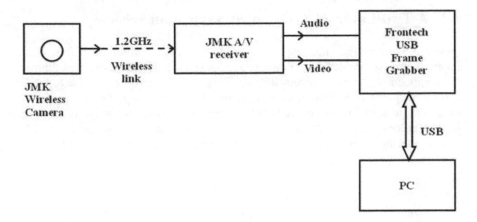

Fig. 3.1(b). The block diagram of the vision system

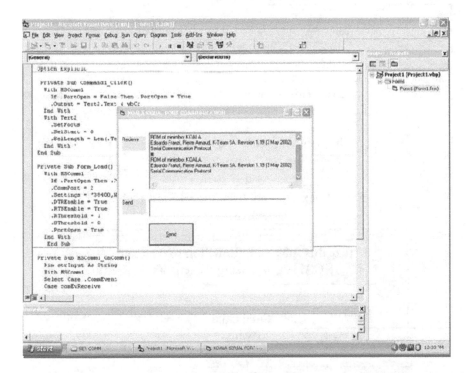

Fig. 3.2. Snapshot of the user-interface developed, that can interact with the user

3.3 A Two-Layer, Goal Oriented Navigation Scheme

Figure 3.3 shows the complete proposed navigation algorithm in a flow chart form. A wireless camera, as shown in Fig. 3.1(a), is used to capture a running video stream of the environment in front of the KOALA robot. An image frame can be acquired from this video stream for further processing at any point of time. This acquired image frame is first processed to make the image suitable for further processing, by employing a series of image processing operations like image filtering, edge detection and image segmentation. Then the shortest path

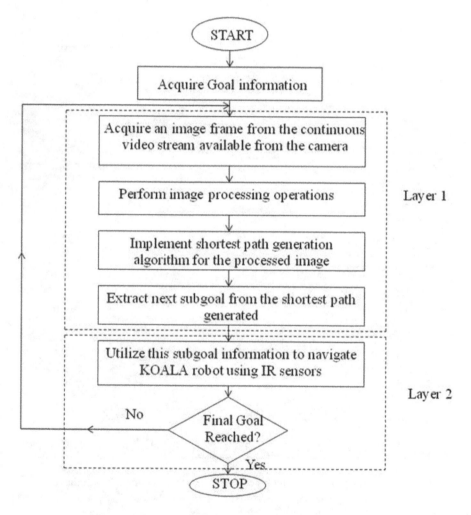

Fig. 3.3. Flow chart for the proposed navigation algorithm

generation algorithm is employed for this processed image, using the goal information, available *a priori*. Next the shortest path generated is utilized to determine the next subgoal. This entire procedure constitutes layer 1 of the algorithm and is implemented in high level in a PC using Visual Basic (VB) platform. This subgoal information is next transferred to layer 2 where the KOALA robot is actually navigated towards the subgoal using obstacle avoidance capability so that the robot can be useful even in a dynamically changing environment. The navigation in layer 2 is performed using several IR sensors, connected at the front face and side faces of the KOALA robot. Once the subgoal is reached, the control is transferred back to layer 1 so that the next subgoal can be generated and actual navigation can be performed in layer 2. This process of local path planning, followed by actual navigation, is continued in an iterative fashion, until the final goal is reached. The algorithm in layer 2 for actual navigation is implemented by developing a C program whose cross-compiled version (a .s37 file) is downloaded in the Motorola processor of the KOALA robot. This .s37 program communicates with the VB program in the PC end, in the interrupt driven mode, in real life. The .s37 program generated from the C program written, is also equipped with the facility of providing support from VB based PC end for a pool of protocol commands for commanding the KOALA robot. These commands are originally only available for execution from a terminal emulator available with the KOALA robot package. We developed a system where all the KOALA robot protocol commands and our additional navigation algorithms are supported by the C program developed, so that the entire system can be completely controlled from the VB platform in the PC end.

3.4   Image Processing Based Exploration of the Environment in Layer 1

Image processing is a form of signal processing where the input signals are images and the output could be a transformed version of the input. The proposed system employs a map building method based on image segmentation, for vision based navigation for mobile robot in an indoor environment, with the assumption that the surface is uniform. The following steps are implemented as follows [27]:

A. Acquire the image from the wireless camera
The camera, mounted at the center of the pan-tilt system of the robot, keeps acquiring a running video stream of the environment ahead of it. From this video stream, an image frame can be acquired for further processing. Figure 3.4(a) shows such an acquired image.

B. Employ low-pass filtering on the acquired image
The acquired image is then low pass filtered to reduce noise. This causes a smoothing or blurring effect on the neighboring pixels. The system is developed using the popular arithmetic mean filter to perform low pass filtering. This arithmetic mean filter is utilized using a 5×5 matrix, centered on each pixel, whose

intensity is computed as the average value of the pixels under the influence of the filter matrix.

C. Detect edges in the filtered image by Canny edge detection

An edge physically signifies a boundary between two regions with relatively distinct gray-level properties. The technique of edge-based segmentation signifies isolation of desired objects from a scene using different types of gradient operators. Edges of the image in our work are detected by using canny edge detection method. Figure 3.4(b) shows the edge image of the processed filtered version of the acquired image.

D. Process the edge image to thicken and link the edges

The edge image contains many small broken edges. To make any edge image a meaningful one, one needs to link nearby edges to bridge gaps and they can be thickened to make their presence distinct. Thickening can be performed by a morphological operation called dilation by a structuring element that is used to grow selected regions of foreground pixels in images. Dilation is normally applied to binary images, and it produces another binary image as output. This dilation operation "thickens" or "grows" objects in a binary image and the shape of thickening can be controlled by a suitable choice of the structuring element shape, used to perform dilation of the image. The concept of linking edges and thickening them by dilation in an edge image can also be performed by a suitable low pass filtering scheme with a suitable choice of the filter mask. This operation is carried out in this work by using geometric mean filtering. The geometric mean filter is member of a set of nonlinear mean filters, which are efficient in removing Gaussian type noise and preserving edge features than the arithmetic mean filter. Figure 3.4(c) shows the edge linked and thickened image.

E. Perform region growing segmentation on the thickened edge image

Once the thickening is done, the image is segregated into regions. To find the obstructed zone and unobstructed zone in the image, region growing based segmentation is performed on the thickened image. Region growing is a simple but efficient region-based image segmentation method and it is classified as one of the pixel-based image segmentation schemes which involves the selection of initial seed points. This approach to segmentation examines the neighboring pixels of the initial "seed points" and determines if the pixel should be added to the seed point or not. Region growing is done by examining properties of each such block created and merging them with adjacent blocks that satisfy some criteria (similar gray-level pixel values, texture etc). The seed point needed for performing region growing is chosen near the bottom center of the image. This point 'S' is shown in Fig. 3.4(c). Now the image is scanned along all the vertical lines from bottom to top. The point at which the floor area ends is regarded as the obstacle. All regions before the obstacles are free zone. All regions beyond the obstacles are termed the hidden zone. Figure 3.4(d) shows the unobstructed zone (free space) with green color and the hidden zone with yellow color. Next the obstructed zone is marked in red and Fig. 3.4(e) shows all these three regions. This entire process is continuous and the obstacle information gets continuously updated.

F. Transform the region grown image to the floor region
The entire grown up region updated with obstacle information is now transformed from image plane to floor region. In order to calculate a distance in the 3D coordinates using single camera, we assume that all the objects have contact at the bottom and interpret it in two dimensional coordinates. Figure 3.5 shows the relationship that, given the elevation of the camera and the elevation angle, how any point on the image plane can be directly mapped on the floor, relative to the position of the camera [25]. Here the robot/camera 3D coordinate frame is assumed with the corresponding notations shown in Fig. 3.5. This coordinate frame is assumed attached to present pose of the robot/camera, at any instant of time. This coordinate transformation mechanism allows one to determine the free points and the obstructed points in the world coordinate system (WCS) from the image acquired by the camera. Hence, with reference to Fig. 3.5, any point with coordinates (u, v) in the image plane can be transformed to the coordinates in the two dimensions (x_c, y_c) in the robot/camera coordinate frame as:

$$x_c = \frac{uh}{\left(f \sin \theta_{EL} + v \cos \theta_{EL}\right)}$$

$$y_c = \frac{h\left(v \sin \theta_{EL} - f \cos \theta_{EL}\right)}{\left(v \cos \theta_{EL} + f \sin \theta_{EL}\right)}$$

where
h = height of the camera optical center from base plane

f = focal length of the camera

θ_{EL} = elevation angle of the camera

$y_{c_{min}} = A(\text{Dead zone})$

$y_{c_{max}} = B$

$x_{c_{min}} = \frac{C}{2}$

$x_{c_{max}} = \frac{D}{2}$

Once this transformation is employed, one can obtain the actual position of a point (x, y) on the floor, given this (x_c, y_c) and the present pose of the robot (x_R, y_R, ϕ_R). Figure 3.4(f) shows the floor with obstacle information. The transformed floor region is in trapezoidal form. Then this floor plane image is copied to the 500 pixel x 500 pixel map which is 20m x 20m as a working space for the robot. Figure 3.6(a) shows a snapshot of the map created and Fig. 3.6(b) shows a snapshot with the floor image in the grid map. In Fig. 3.6(b), the trapezoidal floor region is shown in green color and the obstacle information is shown in red color. The above process of transformation is continuous even when the robot is in motion and it updates the new obstacle information in the map when it is in motion.

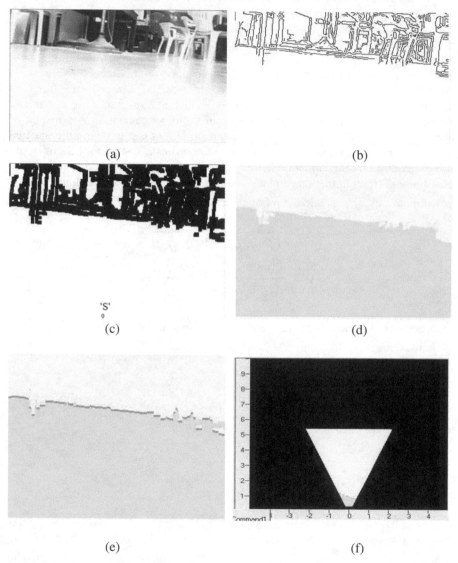

(a)   (b)

(c)   (d)

(e)   (f)

Fig. 3.4. (a) Image acquired by the wireless camera, (b) detected edge image, (c) thickened image, (d) region grown image, (e) image with the obstacle information, and (f) trapezoidal floor image

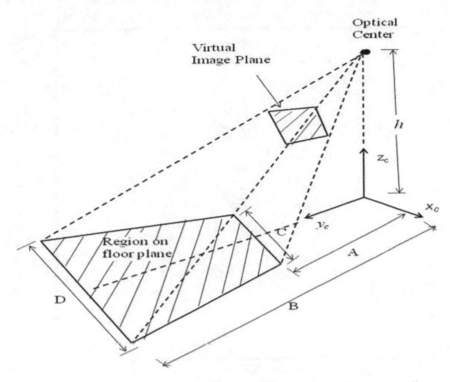

Fig. 3.5. Relationship between the image coordinate and the mobile robot coordinate

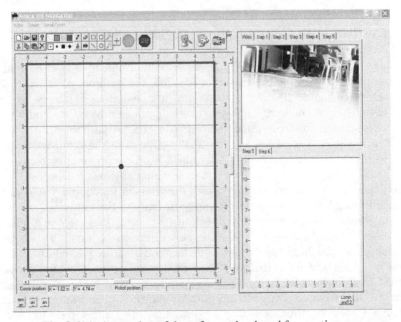

Fig. 3.6(a). A snapshot of the software developed for creating map

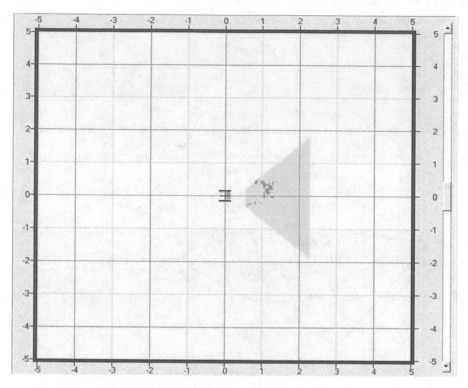

Fig. 3.6(b). A snapshot of the map updated with obstacle information

3.5 Shortest Path Computation and Subgoal Generation

In mobile robot navigation it should be an important objective to determine the optimum path between the present robot location and the goal point, so that the robot can reach the destination in minimum time, avoiding obstacles, as far as practicable. The present work employs a heuristic gradient based method which is based on grid-map for finding the shortest path [26]. Algorithm 3.1 shows this algorithm in detail. The initial and the final positions of the robot are known *a priori* with the obstacle information determined from the previous steps. Now the coordinates along the shortest path are determined by using steepest descent method. The steepest decent algorithm uses the gradient function to determine the direction in which a function is decreasing most rapidly. Each successive iteration of the algorithm moves along this direction for a specified step size and then recomputes the gradient to determine the new direction of travel. This heuristic approach employed here can be easily understood if a land is considered with known obstacles and the initial point and final point on it. The land surface is assumed frictionless such that, say, at the starting point, if we start pouring sand on the ground, it spreads towards all possible paths, similar to dispersion of a fluid in all possible directions. It is obvious that one cannot pass through the obstacles. In each iteration, we assume that a fixed amount of sand is poured and we let it spread. We set a time index to every point on the ground, equal to the iteration number, when the sand reaches a pre-assumed height. So the earlier the height is reached, the smaller is this index. Such a pre assumed height and a fixed amount of sand dispersed are chosen so as to avoid saturation in value within any finite considerable region. Hence a travel time matrix ($\mathbf{H}$) can be calculated employing the finite element diffusion method and this $\mathbf{H}$ matrix is iteratively updated, until a termination criterion is met. At the end of this procedure, those entries in $\mathbf{H}$ which still contain zeros correspond to the obstacle cells. Next, the gradient descent based procedure is employed to determine the coordinates of the points on the shortest path by starting from the goal point and finally arriving at the present robot location. For this, the gradient matrices of $\mathbf{H}$ in x- and y-directions, i.e. $\nabla \mathbf{H}_x$ and $\nabla \mathbf{H}_y$, are calculated and based on them the new co-ordinates of the next point on the shortest path are computed, utilizing the last point obtained on the shortest path. The algorithm always proceeds backwards starting from the goal point. This method is an efficient one and it operates in an iterative fashion. Figure 3.7 shows a sample environment where the shortest path is computed between the initial and the goal point in the map.

BEGIN

1. Obtain the Occupancy grid matrix ($\mathbf{M}$), the start point (x_start, y_start), and the goal point (x_goal, y_goal). $\mathbf{M}(i, j) = 0$ denotes a free cell and $\mathbf{M}(i, j) = 1$ denotes an obstacle cell.

2. Create diffusion matrix ($\mathbf{W}$) and Travel Time Matrix ($\mathbf{H}$) and make them of same size as $\mathbf{M}$.   Initialize $\mathbf{W}_0 = \mathbf{H}_0 = \mathbf{0}$.

3. Set $\mathbf{W}_0 (x\_start, y\_start) = 1$.

4. Set diffusion constant ($diffconst$) and maximum number of iterations without updates (no_update_max). Initialize number of iterations ($iter\_count$) and number of iterations without updates ($no\_update\_iter\_count$).

5. **WHILE** $\left(no\_update\_iter\_count < no\_update\_max\right)$

    5.1. $iter\_count = iter\_count + 1$.

    5.2. Diffuse cells downwards:

    $\mathbf{W}_{iter\_count}(i,j) = \mathbf{W}_{iter\_count}(i,j) + diffconst * \mathbf{W}_{iter\_count}(i+1,j)$
                 $i = 1,2,\ldots,(W\_ROWS - 1); j = 1,2,\ldots,W\_COLS$;

    5.3. Diffuse cells upwards:

    $\mathbf{W}_{iter\_count}(i,j) = \mathbf{W}_{iter\_count}(i,j) + diffconst * \mathbf{W}_{iter\_count}(i-1,j)$
                 $i = 2,3,\ldots,W\_ROWS; j = 1,2,\ldots,W\_COLS$;

    5.4. Diffuse cells towards right:

    $\mathbf{W}_{iter\_count}(i,j) = \mathbf{W}_{iter\_count}(i,j) + diffconst * \mathbf{W}_{iter\_count}(i,j+1)$
                 $i = 1,2,\ldots,W\_ROWS; j = 1,2,\ldots,(W\_COLS-1)$;

    5.5. Diffuse cells towards left:

    $\mathbf{W}_{iter\_count}(i,j) = \mathbf{W}_{iter\_count}(i,j) + diffconst * \mathbf{W}_{iter\_count}(i,j-1)$
                 $i = 1,2,\ldots,W\_ROWS; j = 2,3,\ldots,W\_COLS$;

5.6. Make $\mathbf{W}_{iter\_count}(i,j) = 0$, if $\mathbf{M}(i,j) = 1$;

$\qquad i = 1,2,\ldots,W\_ROWS; j = 1,2,\ldots,W\_COLS$;

5.7. If any $\mathbf{W}_{iter\_count}(i,j)$ becomes greater than the height for the first time, then make corresponding $\mathbf{H}_{iter\_count}(i,j) = iter\_count$.

5.8. Count $sum\_count_{iter\_count}$ as the sum of those entries in $\mathbf{W}$ matrix at present with value > 1.

5.9. **IF** $\left| sum\_count_{iter\_count} - sum\_count_{(iter\_count-1)} \right| < 1$ **THEN**

$\qquad no\_update\_iter\_count = no\_update\_iter\_count + 1$

$\qquad$ **ENDIF**

ENDWHILE

6. All $\mathbf{H}(i,j)$ point still equal to zero are the obstacle points. Set these points to a high value i.e. one more than their adjacent neighbor which one have the highest value (steep gradient for obstacle occupied points).

7. Create shortest path coordinate vectors **sh_path_co   ord_row**   and **sh_path_co   ord_col** and   initialize   the   first   point: **sh_path_co   ord_row** $(1) = x\_goal$; **sh_path_co   ord_col** $(1) = y\_goal$. Set μ.

8. Compute gradient matrices of $\mathbf{H}$ matrix in x-direction $(\nabla\mathbf{H}_x)$ and y- direction $(\nabla\mathbf{H}_y)$.

9. $\nabla\mathbf{H}_x = -\nabla\mathbf{H}_x$; $\nabla\mathbf{H}_y = -\nabla\mathbf{H}_y$; $path\_index =1$; $path\_flag =1$;

10. **WHILE** $(path\_flag =1)$

$\qquad$ 10.1. Compute del_row by interpolation using the $\nabla\mathbf{H}_y$ matrix.

$\qquad$ 10.2. Compute del_col by interpolation using the $\nabla\mathbf{H}_x$ matrix.

$\qquad$ 10.3. Compute the coordinates of the next point on the shortest path:

$\qquad$ **sh_path_coord_row** $(path\_index+1) =$ **sh_path_coord_row** $(path\_index)+$

$$\mu * \frac{del\_row}{\sqrt{del\_row^2 + del\_col^2}}$$

$\qquad$ **sh_path_co ord_col** $(path\_index+1) =$ **sh_path_co ord_col** $(path\_index)+$

$$\mu * \frac{del\_col}{\sqrt{del\_row^2 + del\_col^2}}$$

$\qquad$ 10.4. **IF** (initial point is reached) **THEN**

$\qquad$ $path\_flag = 0$;

$\qquad$ **ENDIF**

ENDWHILE

11. Reverse vectors **sh_path_coord_row** and **sh_path_coord_col.**

END

Algorithm 3.1. The shortest path generation algorithm employing obstacle avoidance

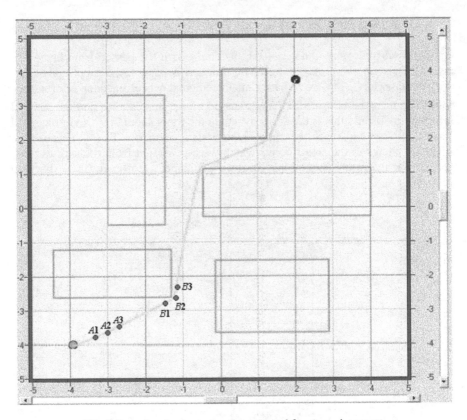

Fig. 3.7. A sample shortest path computed for an environment

Once the shortest path is determined, we need to find the corner points nearer to an obstacle. To find the corner points, we take three consecutive points on the path and find the cosine of the angle between the two line segments joining the first two and last two points. If this value falls below a given threshold, then the middle of these three points is considered as a corner point, otherwise we move to the next subsequent point and again compute the cosine of the new angle. This process is continued until the suitable corner point is obtained. This corner point is stored as the next subgoal point for navigation. For example, in Fig. 3.7, when $A1$, $A2$, $A3$ are the three points under consideration, then the cosine of the angle between the line segments $\overline{A\,1\,A\,2}$ and $\overline{A\,2\,A\,3}$ is very high (above the chosen threshold). So $A2$ is not considered as a corner point. In this process, we keep moving forward, and when we reach the three consecutive points $B1$, $B2$, $B3$, the angle between the line segments $\overline{B\,1\,B\,2}$ and $\overline{B\,2\,B\,3}$ is large enough so that the cosine of the angle is below the chosen threshold. Then $B2$ is considered as a corner point.

3.6 IR Based Navigation in Layer 2

Once the subgoal point is determined, the control will be passed from layer 1 to layer 2. As soon as the new subgoal information is passed, the robot updates its present pose (x_R, y_R, ϕ_R), based on incremental wheel encoder information, and determines the new steering angle, based on its present pose and the subgoal information. Ideally this is the angle by which the robot should turn and proceed at a constant speed to reach the subgoal, in a static scenario. This is because the subgoal belongs to the set of points which were generated from the shortest path generation algorithm, employing obstacle avoidance. However, in a dynamic

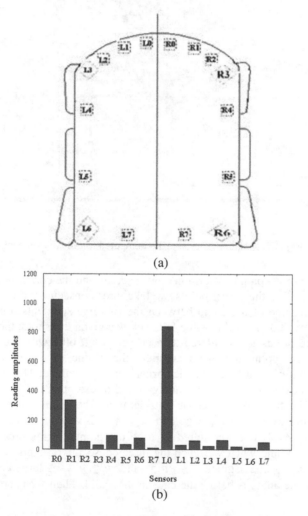

(a)

(b)

Fig. 3.8(a). IR sensor arrangements of the KOALA robot [23]
Fig. 3.8(b). Measured values of the IR sensor readings, by placing a 1.5 cm wide obstacle in front of sensor (R0) at a distance of 10 cm.

scenario, after the last time the vision based mapping subroutine was activated, a new obstacle may have arrived or an old obstacle's position may have been changed. This may result in obstruction along the ideal path of travel between the robot and the subgoal. To cope with this dynamic environment, the navigation is guided by 16 IR sensors, mounted symmetrically along the periphery of the KOALA robot.

These IR sensors are densely populated in front and sparsely populated at the two sides of the robots. Figure 3.8(a) shows the sensor arrangement of the mobile robot and Fig. 3.8(b) shows a typical situation for the measured values of the sensors, by placing a 1.5 cm wide obstacle in front of the front sensor (R0) at a distance of 10 cm from the robot front face. For navigation, these 16 IR sensors scan the environment. Depending on these sensor readings, the system calculates the obstacle regions and free regions ahead of the robot. From these calculations the traversable area is determined. For determining the traversable area, separate thresholds are set for each of the 16 sensors, with the maximum priority given to the front sensors (R0 – R3, L0 – L3). For each sensor, if its reading exceeds its threshold, it means the direction ahead of it is obstructed, else the direction ahead is considered free for traversal. Now, depending on these readings, there can be traversable areas both to the left and to the right of the present pose of the robot. The decision of whether the robot should turn left or right is taken based on which direction will mean that the robot has to undertake the shorter detour with respect to its ideal direction of travel. Once the detour direction is determined, the speed of the robot is determined based on the IR sensor readings in that direction. When the robot travels a predetermined distance, the entire IR based scanning and determination of the new detour direction of traversal is reactivated and this procedure is continued until the robot reaches the subgoal or its closest vicinity. Then the robot stops and the control is transferred back to layer 1.

3.7 Real-Life Performance Evaluation

The performance evaluation has been carried out, for vision based navigation, in our laboratory, utilizing several environments. Here we present the results for four such experiments, two each in static and dynamic environments.

Case Study – I
The initial pose of the robot is (0, 0, 0) and the goal point is (2, 0). There lies an object between the robot and the goal position. It should be mentioned here that for the robot system which is equipped with a pan-tilt mechanism with its corresponding degrees of freedom, in this work, the pan angle and the tilt angle are suitably initialized for a particular environment and then they are kept fixed, for all subsequent experiments. Initially these two angles are so chosen for the robot system developed so that the monocular camera, in each frame, covers a reasonably large floor and environment area. The system is hence equipped with the flexibility where these angles can be suitably initialized depending on the environment where this navigation system is going to be implemented. Figure 3.9(a) shows the image frame acquired from the video stream of the camera

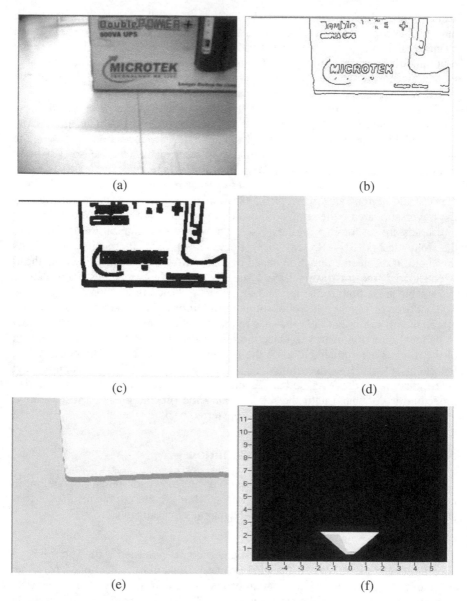

Fig. 3.9. (a): The image acquired, (b)-(f): sequence of image processing caried out in layer 1. (b): edge image; (c): thickened edge image; (d): region grown image; (e): image with free, obstacle, and hidden regions and (f): trapezoidal floor image.

and Figs. 3.9(b)–3.9(f) show the sequence of image processing steps, when the robot is in initial position. The edge of the face of the obstacle on the ground, viewed by the robot in front of it when the robot is at its initial pose, actually extends from (0.9, 0.3) to (0.9, -0.85). Figure 3.10 shows the snapshot of the grid

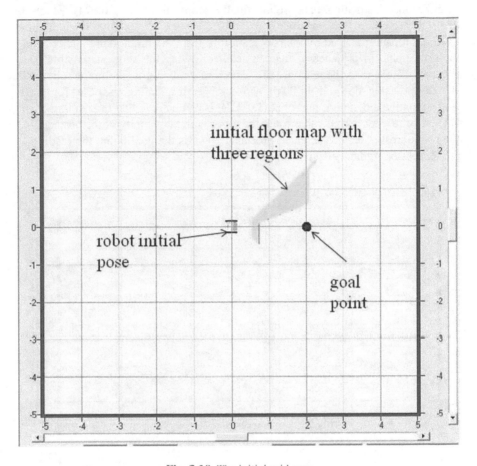

Fig. 3.10. The initial grid map

map with the obstacle information and free region. Here the shortest path is
calculated and the layer 2 of the robot navigation algorithm is updated with the
subgoal information. The algorithm calculates the subgoal 1 as (0.81, 0.33). When
the control is transferred to layer 2, the robot navigates using IR sensor based
guidance, upto subgoal 1. The robot actually stops at (0.819, 0.332) which has
very small discrepancy with the calculated subgoal. Figure 3.11 shows the
snapshot of the grid map when the robot reaches the first subgoal point. This grid
map is developed when the control is transferred back to layer 1 and vision based
processing is carried out once more. Figures 3.12(a)-3.12(d) show the results of
image processing steps, when the robot is at the first subgoal point. These results
are used for IR based navigation once more. This sequential process is continued
to reach the final goal point. Figure 3.13 shows the grid map when the robot

reached the destination. The robot finally stops at (1.963, 0.024) which is extremely close to the specified goal (2,0). Figure 3.14 shows the complete navigation path traversed by the robot, starting from the initial point and reaching the goal point, in presence of the obstacle, following the shortest possible path. Figure 3.15(a) shows the response of IR sensors on the right side of the robot (R0, R3) during navigation and fig. 15(b) shows the corresponding responses for the IR sensors on the left side of the robot (L0, L3). It can be seen that the reading of the R3 sensor reaches a high value when the robot is in the vicinity of the obstacle. As the robot crosses the obstacle and proceeds towards the goal point, the reading of the R3 sensor gradually decreases.

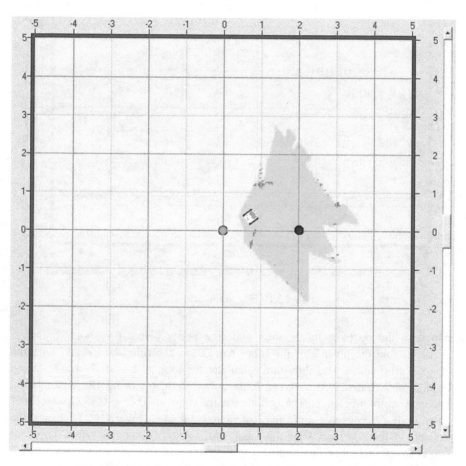

Fig. 3.11. The grid map, when the robot reaches the first subgoal

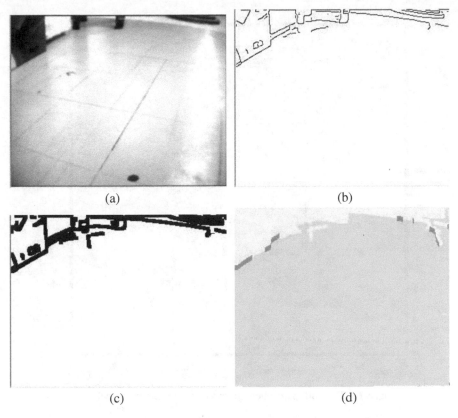

Fig. 3.12. (a)-(d). Results of image processing at subgoal 1. (a): the acquired image; (b): edge image; (c): thickened edge image, and (d): image with three distinct regions.

Case Study – II

Here again the initial pose of the robot is (0, 0, 0) and the new goal point is (3, 0). Now two objects are introduced between the robot and the goal position. Figure 3.16(a) shows the image frame acquired at the initial position of the robot and Figs. 3.16(b)–3.16(f) show the results of subsequent image processing steps in layer 1. Figure 3.17 shows the snapshot of the initial grid map with the obstacle information and free region. Next the shortest path is calculated and the layer 2 of the robot navigation algorithm is implemented with this subgoal information. Figure 3.18 shows the snapshot of the grid map when the robot reaches the first subgoal point. When the robot reaches subgoal 1, the control is transferred back to layer 1. The system again performs the vision based processing, as shown in Fig. 3.19 and Fig. 3.20 shows the grid map when the robot reaches subgoal 2, using IR

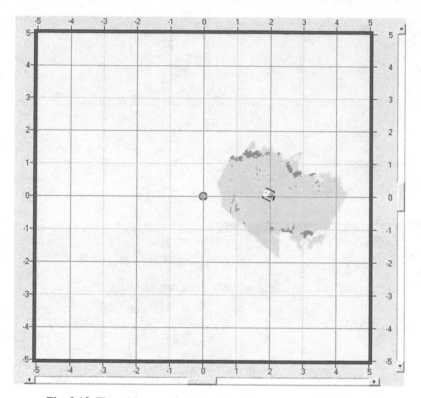

Fig. 3.13. The grid map, when the robot reaches the final goal point

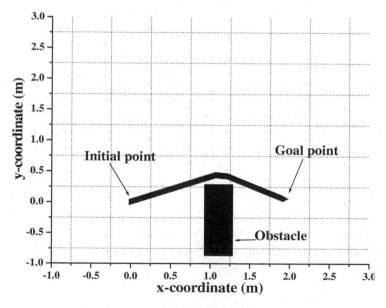

Fig. 3.14. The robot navigation path traversed

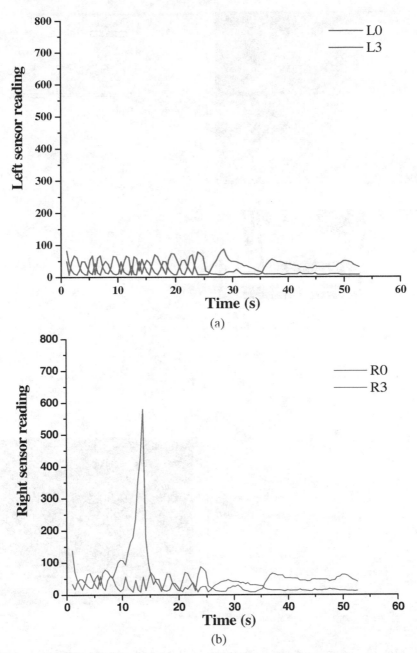

Fig. 3.15. Variation of (a) response of L0 & L3 IR sensors with time and (b) response of R0 & R3 IR sensors with time, for case study I

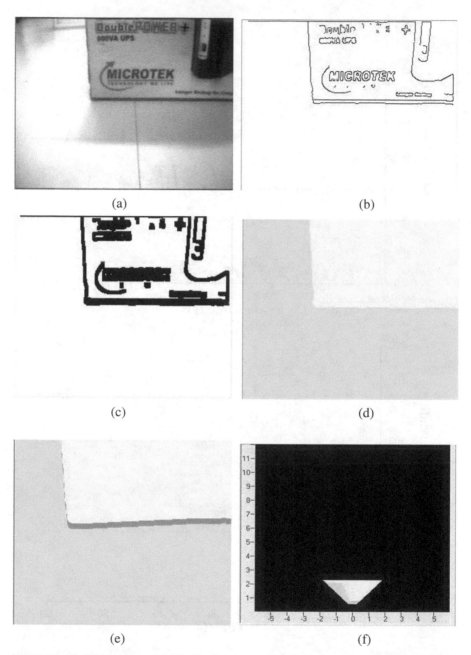

(a)

(b)

(c)

(d)

(e)

(f)

Fig. 3.16. (a): The image acquired, (b)-(f): results of image processing steps in layer 1. (b): edge image; (c): thickened edge image; (d): region grown image; (e): image with free, obstacle and hidden regions and (f): trapezoidal floor image.

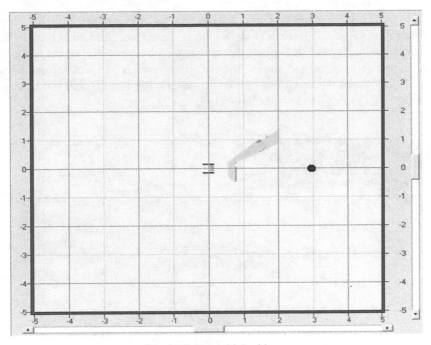

Fig. 3.17. The initial grid map

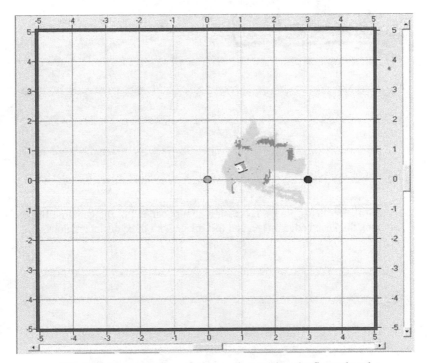

Fig. 3.18. The grid map, when the robot reaches the first subgoal

based navigation in layer 2. This iterative process is continued until the robot reaches the final goal. Figure 3.21 shows the grid map when the robot reaches the final goal point. Figure 3.22 shows the complete path of traversal of the robot for this static environment and shows that the robot reaches the goal satisfactorily. Figure 3.23 shows the variations of four IR sensors, R0, R3, L0, and L3, readings when the robot navigates towards its destination.

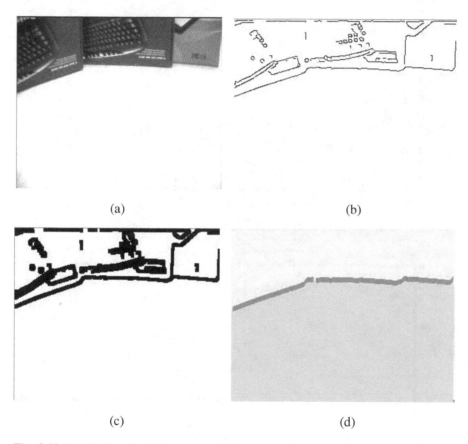

(a)   (b)

(c)   (d)

Fig. 3.19 (a)-(d). Results of image processing at subgoal 1. (a): the captured image; (b): edge image; (c): thickened edge image, and (d): image with three distinct regions.

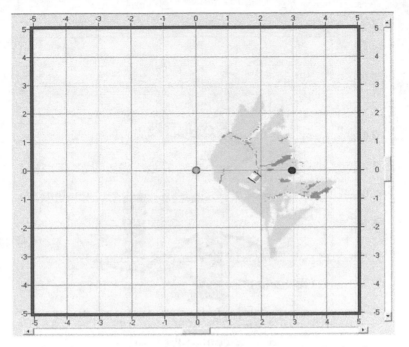

Fig. 3.20. The grid map, when the robot reaches the second subgoal

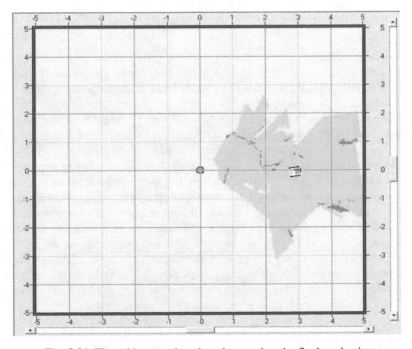

Fig. 3.21. The grid map, when the robot reaches the final goal point

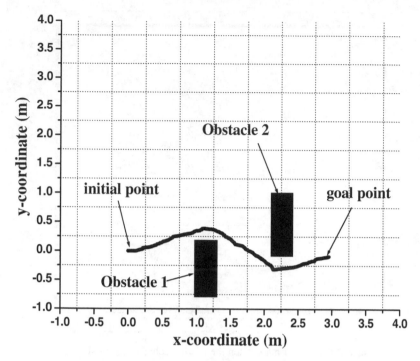

Fig. 3.22. The robot navigation path traversed, in case study I

Case Study – III

In the next two case studies, we demonstrate the utility of the proposed system in case of a dynamically changing environment. Here, for an environment similar to that considered in case study I, the robot starts from an initial pose (0, 0, 0), with a bid to reach the goal point (2,0), in presence of an obstacle between the robot and the goal position. However, after the robot starts its IR based navigation towards subgoal 1, determined using vision based image processing in layer 1 at the initial position of the robot, followed by the determination of the subgoal 1 utilizing the shortest path algorithm, the position of obstacle 1 is shifted. The new position of the obstacle is now shown in Fig. 3.24 where it is moved nearer to the robot and it is shifted towards the left of the robot, with reference to its initial pose. Because of this dynamic variation in the environment, the robot takes a detour towards its left but was still able to avoid the obstacle and reach its subgoal. The subsequent activations of the iterative algorithm show that the robot reaches its final goal almost perfectly, once more. Figure 3.24 shows this navigation of the robot in the dynamic environment. Figure 3.25(a) and Fig. 3.25(b) show the IR sensor readings, in front of the robot. It can be seen that the reading of R0 and L0 receive a sudden kick when the obstacle is moved in the dynamic environment.

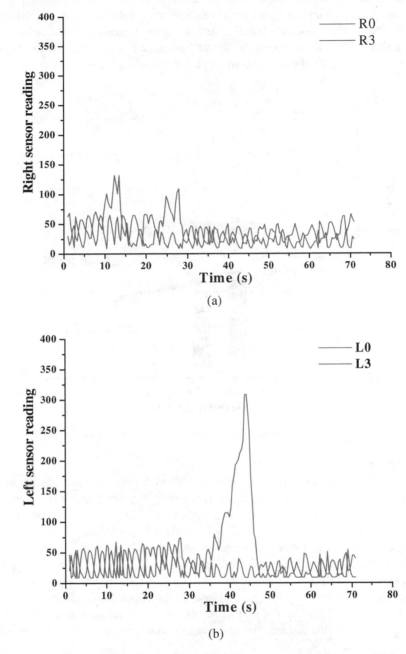

Fig. 3.23. Variation of (a) response of R0 and R3 IR sensors and (b) response of L0 and L3 IR sensors during navigation, for case study II

Here it should be mentioned that if there arises an exceptional situation where the dynamically changing object arrives exactly on a subgoal, then, according to the algorithm, the IR-sensor based actual navigation guidance mechanism will ensure that the robot will stop at the shortest distance from the subgoal, satisfying obstacle avoidance or collision requirement.

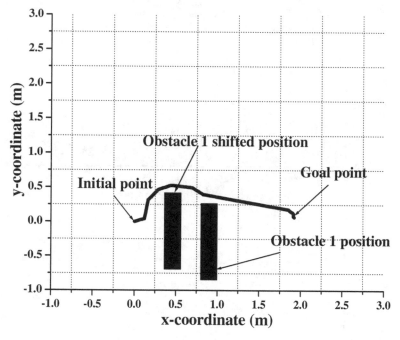

Fig. 3.24. The robot navigation path traversed in a dynamic environment

Case Study – IV

This situation is similar to case study II, but with both obstacles being made dynamic in nature. Here also, after the robot start traversing towards subgoal 1, avoiding obstacle 1 whose position was determined from the vision based image processing in layer 1, the position of the obstacle 1 was suddenly changed. It was brought closer to the robot and more towards its left, making partial dynamic blockage of the free region of traversal. Similarly, when the robot was attempting to traverse a shortest path avoiding obstacle 2, suddenly the position of the obstacle 2 was changed by bringing it closer to the robot. However the robot was able to undertake the required detour in its IR based navigation in each such situation and was able to reach the final goal satisfactorily, as shown in Fig. 3.26.

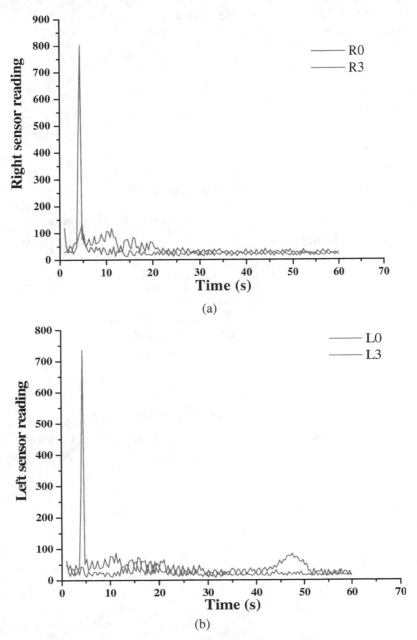

Fig. 3.25. Variation of (a) response of R0 and R3 IR sensors and (b) response of L0 and L3 IR sensors during navigation in the dynamic environment, for case study III

Figure 3.27(a) and 3.27(b) show the readings of the IR sensors R0, R3, L0, and L3. It can be seen that here also the readings of L0 and R0 receive two sudden kicks, when the two obstacle positions are changed.

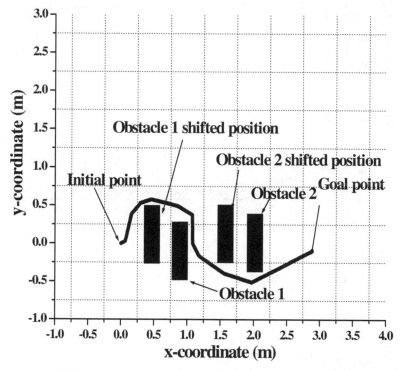

Fig. 3.26. The robot navigation path traversed in a dynamic environment

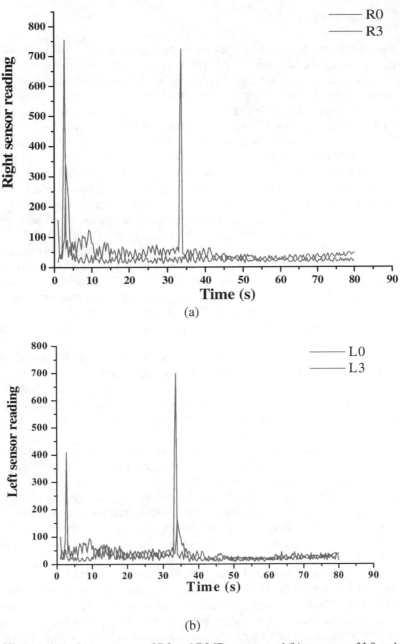

Fig. 3.27. Variation of (a) response of R0 and R3 IR sensors and (b) response of L0 and L3 IR sensors during navigation in the dynamic environment, for case study IV

3.8 Summary

In this chapter we described how a two-layered, goal oriented, vision based robot navigation scheme can be developed. The system employs vision based analysis of the environment in layer 1, which employs several image processing functions and a shortest path generation algorithm, to determine the next subgoal for navigation, with the objective of reaching the final destination as fast as possible, avoiding obstacles. This subgoal information is utilized by the robot in layer 2 to navigate in dynamic environments, utilizing a set of IR sensors, avoiding obstacles, to reach the subgoal or its closest vicinity. This two-layered algorithm is utilized iteratively to create the next subgoal and navigate upto it, so that the final goal is reached sufficiently quickly. This chapter has showed a successful implementation of how to hybridize the shortest path algorithm with camera based image processing to enhance the quality of vision based navigation of mobile robots in the real world, so that, the robot can reach its goal (known *a priori*), following the shortest practical path, avoiding obstacles. The robustness of the system is further ensured by the IR-sensor guided navigation, which helps the robot to adapt its navigation, based on any possible change in obstacle positions in a dynamic environment. This algorithm is implemented for several environments created for indoor navigation in our laboratory. It has been demonstrated that the KOALA robot could achieve its task, each time, satisfactorily, for both static environments and dynamic environments.

The developed programs comprise high-end programs developed in VB platform which communicate, in real-time, with the processor of the robot system, where cross-compiled versions of custom-designed C programs are downloaded. However, in the real implementation phase, the entire system is run from the high-end VB platform in a PC through a user-friendly GUI developed, so that it can be easily utilized by some common users.

For high illumination situations the algorithm is expected and has been demonstrated to provide satisfactory performance. However, for low illumination situations, the reflections of the obstacles on the floor may look dark enough (as is shown in the case of Figs. 3.4(a)-3.4(e)) so that the edge image may contain some edges corresponding to reflections on the floor. Hence these reflections may be interpreted as obstacle and this reduces the free zone computed. However, according to the algorithm, in these exceptional cases, the shortest path computed may be a little longer than the true shortest path but still safe and robust navigation of the robot avoiding obstacles towards the goal will be ensured.

The present system is developed where the robot pose in real environment is estimated by odometry using only incremental wheel encoder information. This suffices well for indoor applications with uniform floors, for which the system is primarily developed. The experiments conducted sufficiently demonstrate that the robot reaches the goal in real world, under these conditions, for a variety of environmental configurations. However, the accuracy of this system may suffer in outdoor environments due to problems like wheel slippage etc. One can undertake such future works into consideration which will attempt to adapt this system for outdoor environments too and this may be accomplished by additionally integrating e.g. extended Kalman filter based algorithms for robot localization, along with the current system developed.

Acknowledgement. The work described in this chapter was supported by University Grants Commission, India under Major Research Project Scheme (Grant No. 32-118/2006(SR)).

References

[1] Chen, Z., Birchfield, S.T.: Qualitative vision-based mobile robot navigation. In: Proc. of the IEEE International Conference on Robotics and Automation (ICRA), Orlando, Florida (May 2006)

[2] Bertozzi, M., Broggi, A., Fascioli, A.: Vision-based intelligent vehicles: state of the art and perspectives. Robotics and Autonomous Systems 32, 1–16 (2000)

[3] Shin, D.H., Singh, S.: Path generation for robot vehicles using composite clothoid segments. The Robotics Institute, Internal Report CMU-RI-TR-90-31, Carnegie-Mellon University (1990)

[4] DeSouza, G.N., Kak, A.C.: Vision for mobile robot navigation: A Survey. IEEE Transactions on Pattern Analysis and Machine Intelligence 24(2), 237–267 (2002)

[5] Moravec, H.P.: Sensor fusion in certainty grids for mobile robots. Artificial Inteligence Magazine 9, 61–74 (1988)

[6] Elfes, A.: Sonar-based real-world mapping and navigation. IEEE Journal of Robotics and Automation 3(6), 249–265 (1987)

[7] Jensfelt, P., Christensen, H.I.: Pose tracking using laser scanning and minimalistic environmental models. IEEE Transactions on Robotics and Automation 17, 138–147 (2001)

[8] Yamauchi, B., Beer, R.: Spatial Learning for navigation in dynamic environments. IEEE Transactions on System, Man, and Cybernetics, Part B 26(3), 634–648 (1995)

[9] Pierce, D., Kuipers, B.: Learning to explore and build maps. In: Proc. of the Twelfth National Conference on Artificial Intelligence, Menlo Park, pp. 1264–1271. AAAI, AAAI Press/MIT Press (July 1994)

[10] Thrun, S.: Learning metric-topological maps for indoor mobile robot navigation. Artificial Intelligence 99, 21–71 (1998)

[11] Santos-Victor, J., Sandini, G., Curotto, F., Garibaldi, S.: Divergent stereo for robot navigation: Learning from Bees. In: Proc. IEEE Computer Society Conference on Computer Vision and Pattern Recognition, New York, pp. 434–439 (June 1993)

[12] Kim, D., Nevatia, R.: Recognition and localization of generic objects for indoor navigation using functionality. Image and Vision Computing 16(11), 729–743 (1998)

[13] Murray, D., Little, J.J.: Using real-time stereo vision for mobile robot navigation. Autonomous Robots 8, 161–171 (2000)

[14] Davison, A.J.: Mobile robot navigation using active vision. PhD thesis (1998)

[15] Ayache, N., Faugeras, O.D.: Maintaining representations of the environment of a mobile robot. IEEE Transactions on Robotics and Automation 5(6), 804–819 (1989)

[16] Fialaa, M., Basub, A.: Robot navigation using panoramic tracking. Pattern Recognition 37, 2195–2215 (2004)

[17] Gasper, J., Santos- Victor, J.: Vision-based navigation and environmental representations with an omnidirectional camera. IEEE Transactions on Robotics and Automation 16(6), 890–898 (2000)

[18] Ohya, A., Kosaka, A., Kak, A.: Vision-based navigation by a mobile robot with obstacle avoidance using single-camera vision and ultrasonic sensing. IEEE Transactions on Robotics and Automation 14(6), 969–978 (1998)

[19] Li, M.H., Hong, B.H., Cai, Z.S., Piao, S.H., Huang, Q.C.: Novel indoor mobile robot navigation using monocular vision. Engineering Applications of Artificial Intelligence, 1–18 (2007)

[20] Latombe, J.: Robot Motion Planning. Kulwer, Norwell (1991)

[21] Dijkstra, E.W.: A note on two problems in connection with graphs. Numerische Mathematik 1, 269–271 (1959)

[22] Nilsson, N.J.: Principles of Artificial Intelligence. Tioga Publishing Company (1980)

[23] KOALA User Manual, Version 2.0(silver edition), K-team S.A., Switzerland (2001)

[24] Singh, N.N., Chatterjee, A., Rakshit, A.: A PIC microcontroller-based system for real-life interfacing of external peripherals with a mobile robot. International Journal of Electronics 97(2), 139–161 (2010)

[25] Kim, P.G., Park, C.G., Jong, Y.H., Yun, J.H., Mo, E.J., Kim, C.S., Jie, M.S., Hwang, S.C., Lee, K.W.: Obstacle Avoidance of a Mobile Robot Using Vision System and Ultrasonic Sensor. In: Huang, D.-S., Heutte, L., Loog, M. (eds.) ICIC 2007. LNCS, vol. 4681, pp. 545–553. Springer, Heidelberg (2007), doi:10.1007/978-3-540-74171-8.

[26] http://www.mathworks.com/matlabcentral/fileexchange/8625-shortest-path-with-obstacle-avoidance-ver-1-3

[27] http://atalasoft-imgx-controls-sdk.software.informer.com/

[28] Nirmal Singh, N., Chatterjee, A., Chatterjee, A., Rakshit, A.: A two-layered subgoal based mobile robot navigation algorithm with vision system and IR sensors. Measurement 44(4), 620–641 (2011)

[29] Nirmal Singh, N.: Vision Based Autonomous Navigation of Mobile Robots. Ph.D. Thesis, Jadavpur University, Kolkata, India (2010)

Chapter 4
Indigenous Development of Vision-Based Mobile Robots

Abstract. In this chapter we shall discuss how a low-cost robot can be indigenously developed in the laboratory with special functionalities. Especially, the development of two types of PIC microcontroller based sensor systems that can be integrated with a robot will be discussed in detail in this regard. One of them will be the development of an IR range finder system that can be developed with dynamic range enhancement capability. The second one will be the development of an optical proximity detector system which utilizes the principle of switching mode synchronous detection technique.

4.1 Introduction

In the phase of implementing the vision based algorithms with the KOALA robot (the version of KOALA that was procured by us), it was found that the KOALA robot operates under certain constraints, as given below:

- The communication between PC/Laptop and KOALA robot takes place by means of RS232. However, most of the present day PCs/Laptops do not have any serial interface and hence they require a separate USB-to-serial converter, to operate in conjunction with the KOALA robot.
- For a PC-KOALA combination, high-speed data transfer is not possible.
- KOALA I/O interface is limited.
- KOALA does not have any provision for USB interface.
- Low-cost USB webcam cannot be connected to KOALA directly.
- Image processing cannot be accomplished with KOALA's low-end processor. This requires a separate on-board Laptop or a PC with wireless camera interface. This makes the arrangement become complex and bulky.

Hence, a robot is developed indigenously in our laboratory, with an aim to overcome the above drawbacks and the functionalities and capabilities of this robot are described in detail in this chapter.

A. Chatterjee et al.: Vision Based Autonomous Robot Navigation, SCI 455, pp. 83–100.
springerlink.com

4.2 Development of a Low-Cost Vision Based Mobile Robot

As described in the beginning of this chapter, a mobile robot setup is indigenously developed, with an aim to provide a low-cost solution to the industrial community [14]. Figure 4.1 shows the actual robot in its front view and bottom view. Figure 4.2 shows the block diagram representation of the robot. The robot developed is a two-wheeled, differential drive system. The robot is equipped with six IR proximity sensors, one IR range sensor system, and a laptop. The proximity sensors provide Boolean signals, where each sensor gets activated if the robot is sufficiently close to an obstacle, or remains deactivated otherwise. The IR sensor based system adds a degree of freedom to the system as its angular position is controlled by a servo motor. This enables the IR sensor to scan the front and the side of the robot environment at eleven angular positions, from left to right. A laptop is mounted on the robot system so that the robot becomes a stand-alone, self-sufficient system. The laptop comprises 4GB solid-state HD, 1GB RAM, with Windows XP SP2 operating system. The laptop is free from any moving parts and it communicates with the robot base through a USB link. The robot base is energized (5V, 1A) from the laptop through two USB cables and no separate power source is needed for the mobile robot operation. All the RC servos employed are power controlled for energy saving. The left and right wheel encoders (4-pulses/rotation) are developed using hall-effect switches. The laptop camera with auto-focus serves as the mono-vision sensor of the robot system. The robot uses the webcam of the laptop as its mono-vision sensor. The IR range sensor system is specially developed for obstacle detection and avoidance, which employs a microcontroller (PIC 12F683) based system, also indigenously developed, with an aim to enhance the dynamic range of the range finder system. The system employs a Visual Basic based robot control program and navigation is performed using vision and IR range sensors. The system is also equipped with a Wi-fi link for wireless remote monitoring and supervision.

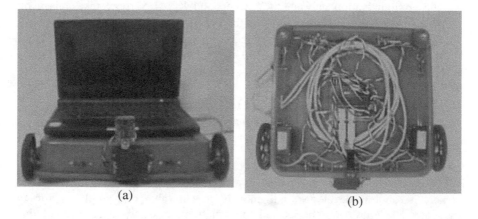

(a)   (b)

Fig. 4.1. The mobile robot, developed indigenously, in its (a) front view and (b) bottom view

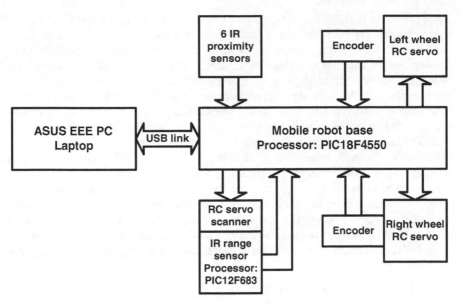

Fig. 4.2. The schematic diagram of the mobile robot

4.3 Development of Microcontroller Based Sensor Systems for Such Robots

This robot developed is made equipped with three special functionalities. The robot comprises two special types of sensor systems developed with indigenous concepts: (a) infrared (IR) sensors with the capability of dynamic range enhancement [2] and (b) optical proximity detectors using switching-mode synchronous detection technique [15]. These sensor systems are developed using PIC microcontrollers. In addition to this, the robot system is equipped with a sophisticated capability of intranet-connectivity where the laptop mounted on the robot, acting in a slave mode, can be suitably commanded by a PC, acting in the master mode, situated in a remote end.

4.3.1 IR Range Finder System with Dynamic Enhancement[1]

The robot system developed is equipped with an indigenously developed PIC Microcontroller based IR range finder system, with dynamic range enhancement capability [2]. Infrared (IR) range finders are overwhelmingly employed in robots for range measurement because of small size, ease of use, low-cost, and low-power consumption. In its conventional form, the Sharp make IR range finder

---

[1] Section 4.3.1 is based on "A microcontroller based IR range finder system with dynamic range enhancement", by Anjan Rakshit and Amitava Chatterjee, which appeared in IEEE Sensors Journal, vol. 10, no. 10, pp. 1635-1636, October 2010. © 2010 IEEE.

finds extensive real-life use, which uses the method of triangulation [1]. Here, the angle of light reflected from the object depends on the object range. In our robot, the IR range finder system employed is developed using scattered radiation-based sensing, which attempts to reduce the influence of orientation of the plane of the object on the sensor reading obtained, as is the case in traditional triangulation-based approach. Usually the output voltage from an IR range finder system increases with decrease in range of the object, i.e. for a nearer object. However, the system can only be used beyond a *dead zone* because, for any range value within this dead zone, the voltage starts decreasing again, instead of increasing [1]. This is because, within the dead zone, probability of the narrow IR beam missing the sensor becomes significant. To increase the sensitivity of the IR sensor based obstacle avoidance scheme, the robot system, instead of utilizing a simple IR range sensor, is built with the PIC microcontroller based IR range finder system, developed in-house [2]. The system developed here utilizes an array-based approach where the burst frequency and duration of IR energy transmitted are progressively reduced. The objective is to reduce the dead zone, by utilizing the

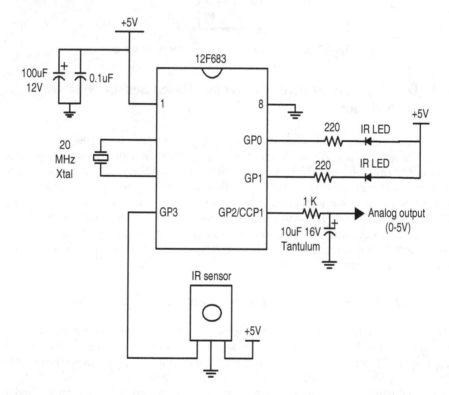

Fig. 4.3. The PIC 12F683 microprocessor based IR range finder system developed, for dynamic range enhancement (Reproduced from [2] with permission from the IEEE. ©2010 IEEE).

output from the IR sensor system to adaptively switch an IR LED ON/OFF. The system employs two IR sources on two sides of the IR sensor whose spatial separation helps to achieve the range enhancement.

Figure 4.3 shows the hardware system developed, in its schematic form, utilizing a PIC12F683 microcontroller [3]. The IR energy transmitted by two high intensity infrared LEDs (IR_LED1 and IR_LED2) is received by a SHARP-make IR sensor system (IS 1U60), called IR_Sensor in Fig. 4.3. The internal block diagram of the IS 1U60 system [4] shows that, when this receiver receives IR energy input, the sensor output goes low and vice versa. The center frequency of the bandpass filter is $f_0 = 38$ kHz. The relative sensitivity is maximum around the carrier frequency of 38 kHz [4], utilized for frequency modulation purpose. In the nominal case, burst wave signals of 38 kHz frequency, with a 50% duty cycle, are transmitted, for a duration of 600 μs [4].

Algo. 4.1 shows the main routine implemented in the PIC microcontroller. Algo. 4.2 shows the real-time interrupt routine developed, enabled on Timer1 overflow, that works in conjunction with the main routine. We introduce two arrays: (i) the *Burst_Freq_Array* for controlling the carrier or burst frequency of IR_LEDs and (ii) the *Integral_Cycle_Array* which determines how long the IR_LEDs should transmit in one sweep. In conventional systems, the burst frequency is 38 kHz, with a 50% duty cycle, the transmission duration is 600 μs, and the sensors produce sensitive results for a narrow width of relatively large ranges. We intentionally manipulate these two variables so that the IR_SENSOR receive some amount of IR light energy, reflected back from the object, for several or a few of these burst frequency durations during one sweep, depending on the distance. This information (*Range_Count*) is exponentially averaged to prepare a steady PWM signal. For a reasonable sensor speed, we can only build these arrays of finite lengths, that gives rise to "range quantization" or finite resolution of the system developed.

BEGIN
1. Initialize IR_LED1 and IR_LED2 in OFF mode.
2. Prepare *Burst_Freq_Array* and *Integral_Cycle_Array*.
3. Prepare Timer1 register pair for Timer1 interrupt.
4. Program suitable PWM carrier frequency.
5. Receive *Range_Count* info. from interrupt routine.
6. Scale this info. suitably for PWM generation.
7. Generate PWM signal using exponential averaging.
8. Go to step 5.
END

Algo. 4.1. Main routine in PIC microcontroller

BEGIN
1.   Prepare for interrupt using *Burst_Freq_Array*[i].
2.   Set *Count1_max = Internal_Cycle_Array*[i].
3.   **IF** (*Count1 > Count1_max*),
              Toggle *Burst_Duration_flag* and Reset *Count1*.
              IF (*Burst_Duration_flag* == 0),
                    Increment *i* by 1.
                    IF (SIGIN == 0),
                          Increment *j* by 1.
                    ENDIF
              ENDIF
              IF (*i* reach last entry in *Burst_Freq_Array*),
                    Range_Count =j; Reset *i* and *j*;
              ENDIF
     ENDIF
4.   **IF** (*Burst_Duration_flag* == 1),
              Put IR_LED1 ON if *Burst_Freq_flag* = 0.
              Put IR_LED2 ON if both *Burst_Freq_flag* = 0 and SIGIN = 0.
                    ELSE
                          Put both IR_LED1 and IR_LED2 OFF.
              ENDIF
5.   Toggle *Burst_Freq_flag*.
6.   Clear *Interrupt_flag*.
END

Algo. 4.2. Interrupt routine

4.3.1.1   The Dynamic Range Enhancement Algorithm

The objective of dynamic range enhancement is achieved by utilizing the output from the IR_SENSOR as a feedback signal (SIGIN) to the microcontroller, which adaptively turns IR_LED2 ON/OFF. Algo. 4.2 shows that the blinking of IR_LED2 is controlled by the states of both *Burst_Freq_flag* and SIGIN. In a conventional IR range finder, within the dead zone, most IR energy reflected back from the object cannot be sensed by the IR_SENSOR. In our system, for distant objects, mostly only IR_LED1 blinks. As we approach the dead zone gradually, IR_LED2 starts getting activated often, as there is a higher probability of SIGIN being low. This intelligent scheme adaptively puts IR_LED2 ON more often with decreasing range, in an intelligent manner, which helps to reduce the length of the dead zone and achieves the required dynamic range enhancement. This is in stark contrast with the working principle of a conventional IR range finder, where, within the dead zone, most of the IR energy reflected back from the object cannot be sensed by the IR_Sensor.

4.3.1.2 Experimental Results

We carried out an experiment in our laboratory, where, for the system without range enhancement, we do not utilize the feedback signal SIGIN to control IR_LED2 in our interrupt routine. Figure 4.4 shows the output voltage vs. range variations for these two cases. For each range/distance, the output voltage computed is the average of ten readings taken, for both with and without range enhancement case. For the system without dynamic range enhancement, the usable range is 25-50 cm and below 25 cm the dead zone arrives. It can be seen that our proposed system could reduce this dead zone and the dynamic range was enhanced with the usable range being 10-50 cm.

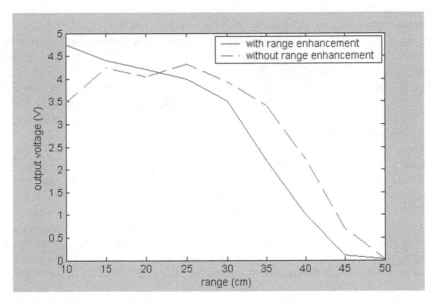

Fig. 4.4. Output voltage vs. range curve for IR sensor system. (Reproduced from [2] with permission from the IEEE. ©2010 IEEE.).

4.3.2 Optical Proximity Detectors Using Switching-Mode Synchronous Detection Technique[2]

The indigenously developed robot system is also equipped with optical proximity detectors which are developed utilizing the theory of switching-mode synchronous detection in a PIC microcontroller based application [15]. Microcontroller based systems have been widely used, in recent times, to develop such low cost robotic

---

[2] Section 4.3.2 is based on "A microcontroller based compensated optical proximity detector employing switching-mode synchronous detection technique", by Anjan Rakshit and Amitava Chatterjee which appeared in Measurement Science and Technology, vol. 23, no. 3, March 2012. Reproduced with kind permission of IOP Publishing Ltd. [Online]: http://m.iopscience.iop.org/0957-0233/23/3/035102

sensors systems [2] and also several intelligent instrumentation systems [5-7]. In this section we describe the development of a PIC microcontroller [3] based optical proximity detection sensor system which is developed using switching mode synchronous detection technique, an efficient strategy used to extract fundamental component of a signal heavily corrupted with noise. Such synchronous detection techniques have been popularly employed in AM radio receivers, in ac-biased strain-gauge bridge circuits, in pyrometer systems [8], in mechanical vibration measurement [9], in synchronous phase to voltage converters [10], in fiber optic sensor-based measurements [11], etc. The objective here is to develop a low cost yet powerful robot sensor that can provide accurate proximity indication of obstacles, even with a wide variation of ambient illumination conditions. This system is developed using two white LEDs which emit light to determine proximity of an obstacle. An electronic circuit using a light dependent resistance (LDR) [12] in conjunction with a transistor determines whether an obstacle is in close enough proximity or not. The system is developed with external threshold variation flexibility so that the maximum obstacle distance causing activation of the sensor can be suitably varied for different working conditions. The sensor system developed has an additional important merit that it has dynamic compensation capability so that the sensor performance is designed to be almost independent of ambient illumination conditions.

There are some important factors that influence the performances of such proximity sensors. The detection of an object will essentially depend on the detection of the radiation reflected back from the surface of the object and, hence, for the same closeness or proximity of an object from the sensor, the amount of radiation reflected back will depend on the reflectivity of the object. The reflectivity of the object varies between 0 and 1. A highly reflecting object will have a reflectivity close to unity and vice versa. Another important factor of influence is the condition of the surface i.e. how smooth (or rough) the surface of the object is on which the light energy from the white LED sources are incident. It is known that, if the reflecting surface is large enough to encompass the entire spatial distribution of the light emitted by the two LEDs, then, for dull objects, the sensor's analog signal can be used to determine the proximity distance, if the surface reflectivity is known. However, in most practical situations, the robot sensor does not know the type of object it is going to encounter during its navigation, and hence, the numerical values of their reflectivities will not be known a priori. To consider such situations, we have conducted experiments for a set of objects having wide variations in reflectivities and hence the suitability of the sensor developed is extensively tested and verified.

4.3.2.1 PIC Microcontroller Based Optical Proximity Detector

Figure 4.5 shows the PIC 12F675 microcontroller based system developed. This system has two digital outputs (pin 3 and 5) connected to two white LED drives (LED1 and LED2), two analog inputs (pin 6 and 7) and one digital output (pin 2) to turn an LED (named PXD_LED) ON/OFF. The pin 7 input is obtained from the collector of a P-N-P transistor whose emitter circuit contains a light dependent resistor (LDR) [12], whose resistance varies with the illumination. The system is designed with an external preset potentiometer at input pin 6, to adjust a threshold voltage (*THLD_val*), essential for preventing any spurious activation of PXD_LED. Each white LED is driven by an identical rectangular pulse, turning them simultaneously ON/OFF for a chosen time duration.

Algo. 4.3 shows the main routine implemented in the PIC 12F675 microcontroller. Algo. 4.4 shows the real-time interrupt routine developed in conjunction with the main routine and it is enabled on Timer1 overflow. The system is so designed that each time interrupt is generated at an interval of 1 ms. At each such interrupt generation, the value of a counter, named as *count1*, is incremented by 1. According to the design philosophy chosen, the ON and OFF durations of the rectangular pulse driving each white LED are unequal and the ON time duration, in each cycle, is controlled by a designer chosen parameter, *Count1_on_max*, and the OFF time duration, in each cycle, is controlled by a designer chosen parameter, *Count1_off_max*. In each cycle, as long as the value of *count1* remains within *Count1_on_max*, both LED1 and LED2 remain ON. For the time duration when the value of *count1* remain within the band [*Count1_on_max*, *Count1_off_max*] both LED1 and LED2 remain OFF. When both these LEDs are ON, they emit optical radiation. The proximity of an object is determined on the basis of the amount of optical radiation reflected back from a nearby object and this is determined in terms of the voltage signal received at input pin 7, from the output of the LDR-transistor combination. For each such acquisition of an input signal, it is always carried out towards the end of the duration of an ON/OFF time period. This is done to allow analog signal stabilization before any measurement is actually carried out. Hence any such signal acquisition is carried out at those instants when (*Count1* == (*Count1_on_max*-2)) or (*Count1* == (*Count1_off_max*-2)).

Each such signal acquired is subjected to three-point median filtering to eliminate any spurious high frequency component, especially impulse natured signals, which may have contaminated the original signal. The signal acquired at pin 7 and then median filtered is called *LDR_on_val*, when this is acquired during ON time of the white LEDs. The identically acquired and processed signal is called *LDR_off_val*, when this is acquired during OFF time of the white LEDs. One can easily appreciate that, if there is a sufficiently close object/obstacle, then *LDR_on_val* will be significantly higher than the *LDR_off_val*. Hence, ideally speaking, a higher value of (*LDR_on_val* - *LDR_off_val*) means a closer object and if this (*LDR_on_val* - *LDR_off_val*) exceed a threshold value then the output PXD_LED will be turned ON, indicating the activation of proximity detection sensor. However, depending on different environments, there are possibilities that, if this threshold value is made a fixed one, then, in certain situations, PXD_LED may get turned ON, even when the object is not in near proximity. Hence, to avoid such spurious activations, the user is given the flexibility where they can externally set a POT using which they can regulate the threshold value chosen

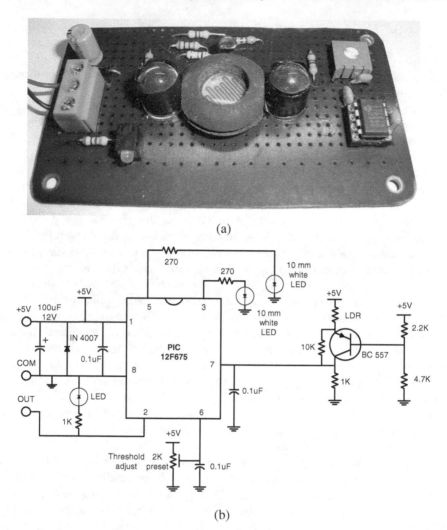

Fig. 4.5. The optical proximity detector system developed: (a) the hardware system and (b) the schematic diagram

(acquired, processed by median filtering and named as *THLD_val*). From Algo. 4.3, if (*LDR_on_val* - *LDR_off_val*) exceed *THLD_val*, then one can conclude that the proximity sensor is close to an obstacle and the output PXD_LED will be turned ON, otherwise it will be OFF.

The developed system also employs a smart compensation scheme that can dynamically cope with ambient illumination variations. The design of the LDR-transistor combination circuit has been so carried out that the transistor always maintains almost constant voltage across the LDR to ensure same signal level, independent of different ambient illumination conditions. Hence an approximately constant signal level is ensured for the two extreme cases of both weak and strong ambient illuminations. This ensures almost linear sensitivity of the sensor i.e. an almost constant ratio of incremental variation in output voltage (i.e. input voltage at pin 7) to the incremental variation in the relative distance between the optical sensor and the obstacle i.e. ($\Delta V / \Delta x$) value. This directly translates into a very important property of any sensor system designed i.e. provision for almost constant detector output voltage variation with the same change in primary measurand (in this case, distance between the sensor and the obstacle), in spite of variation in other secondary factors (in this case, ambient illumination).

BEGIN
1. Prepare Timer1 for 1 ms Timer1 interrupt.
2. **IF** (*Count1* == (*Count1_on_max*-2)),
   Accept the input signal from LDR-transistor combination circuit at pin 7.
   ENDIF
3. **IF** (*Count1* == (*Count1_on_max*-1)),
   Median filter the 10-bit ADC converted analog signal in PIN 7 and store it as *LDR_on_val*.
   ENDIF
4. **IF** ((*Count1* >= *Count1_on_max*) & (*Count1* < *Count2_on_max*)),
   Accept the THLD set as an input signal at pin 6.
   Median filter the 10-bit ADC converted analog signal in PIN 6 and store it as *THLD_val*.
   ENDIF
5. **IF** (*Count1* == (*Count1_off_max*-2)),
   Accept the input signal from LDR-transistor combination circuit at pin 7.
   ENDIF
6. **IF** (*Count1* == (*Count1_off_max*-1)),
   Median filter the 10-bit ADC converted analog signal in PIN 7 and store it as *LDR_off_val*.
   ENDIF
7. **IF** ((*LDR_on_val-LDR_off_val*) > *THLD_val*),
   IF (*Count2* == *Count2_max*),
   Reset *Count2* to 0.
   ENDIF
   IF ((*Count2* > (*Count2_max*-10)) & (*Count2* < *Count2_max*)),
   Turn PXD_LED on.
   ENDIF
   ENDIF
END

Algo. 4.3. Main routine in PIC microcontroller

BEGIN
   1.   Prepare for 1 ms timer interrupt.
   2.   Increment *Count1* by 1.
   3.   **IF** (*Count2* < *Count2_max*),
          Increment *Count2* by 1.
      ENDIF
   4.   **IF** (*Count1* > *Count1_on_max*),
          Turn both LED1 and LED2 off.
      ENDIF
   5.   **IF** (*Count1* > *Count1_of_max*),
          Reset *Count1* to 0.
          Turn both LED1 and LED2 on.
      ENDIF
END

Algo. 4.4. Interrupt routine

4.3.2.2   Switching Mode Synchronous Detection (SMSD) Technique

The synchronous detection is a popular signal processing technique used to extract fundamental component of a weak signal, embedded within a strong noisy counterpart. This technique is popularly employed in radio communication, in industrial scenario (where there is strong possibilities of encountering heavily noise contaminated or disturbed signals) etc. and this technique requires a reference signal with known frequency and phase [8]. A very popular application of synchronous detection technique includes design of superheterodyne receivers for AM radio. In traditional synchronous detection method, the reference signal employed is a pure sinusoidal signal or a harmonic signal. In a popular variation of this traditional technique, switching mode synchronous detection technique employs a square/rectangular wave as a reference signal. The core of a switching mode synchronous detector employs a phase sensitive detector. In SMSD technique [13], a periodic rectangular pulse train is employed as a reference $r(t)$ which is used to sample the noisy signal $x(t)$ and the output of the detector $x_m(t)$ is low pass filtered to recover the fundamental of $x(t)$, i.e. $x_f(t)$. In our scheme, we employ a modified switching mode synchronous detection technique, as shown in Fig. 4.6. Here, the rectangular reference $r(t)$ is used to sample the noisy signal $x(t)$ and this produces the output of the detector $x_m(t)$, identical to a conventional SMSD scheme. Then the output $x_m(t)$ is used to cause an activation of the output LED only when this $x_m(t)$ produces a high signal for a consistently long, continuous duration of time. This is similar to a conditional sample and hold operation and can be visualized equivalent to a low pass filtering action, because it avoids any spurious activation of the proximity detector caused by any high input impulse signal or a short duration input signal, acquired at input pin 7 of the microcontroller, which may have arose because of some unwanted, external interference. If this signal produces a high value for a continuously long time then

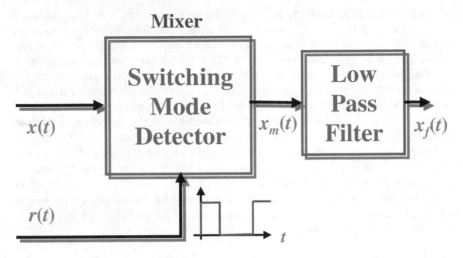

Fig. 4.6. The modified switching mode synchronous detector

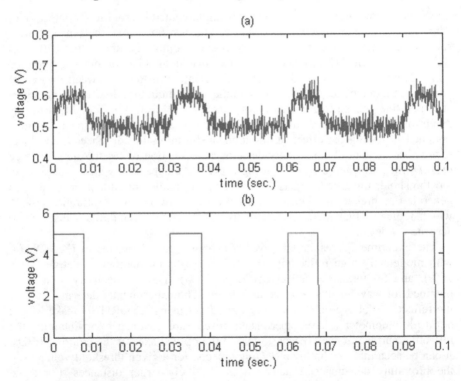

Fig. 4.7. (a) A sample real input signal $x(t)$ and (b) the reference signal $r(t)$

we can infer that it is definitely because of the presence of an object in proximity of the sensor and not because of any noisy signal acquired. Figure 4.7 shows a sample real situation for a given condition of an object in the proximity of the sensor. The input signal acquired at pin 7 of the microcontroller is shown as input $x(t)$ and the reference signal is shown as $r(t)$. It should be borne in mind that, in switching mode synchronous detection technique, the relative phase of the signal under consideration and the reference signal plays an important role [8]. For those frequencies in the signal whose phase do not match with the reference, the output reduces and a given frequency has zero contribution in the output of the switching mode detector, if its phase is at a $90°$ deviation from the reference signal. In our scheme, the transistor emitter signal output read at pin 7 is the signal $x(t)$ and the white LEDs produce the reference signal $r(t)$. In case of sufficient proximity of an object, the low pass filter produces a high output and for distant objects the output is low. The THLD signal is utilized on whose basis the proximity of an object is determined as a Boolean signal.

4.3.2.3 Experimental Results

The optical proximity detector designed is implemented in real life for detection of nearby objects under several case study conditions. Each time the sensor system showed satisfactory performance with a Boolean output i.e. the output LED (i.e. PXD_LED is turned ON for sufficient proximity of an object or, otherwise, turned OFF). However, as remarked earlier, if the relative distance between the sensor and a distant object keeps reducing, then the exact minimum distance of an object at which this change in Boolean output takes place, from OFF condition to ON condition, depends on various factors. Figure 4.8 shows the experimental results obtained in testing the effect of variation in the minimum distance of an object required to activate the proximity detector as a function of the threshold voltage (*THLD_val*), adjusted externally using a POT. As expected, with an increase in the threshold, the detector gets activated for a smaller minimum proximity, in general. For higher thresholds set, the system shows a near saturation effect, which indicates that there is an effective dead zone for minimum distance to activate the detector.

The experimental results are given for three types of objects in Fig. 4.8: (a) with moderately high reflectivity (p = 33%), (b) with medium reflectivity (p = 16%), and (c) with low reflectivity (p = 7.8%). These reflectivity values are obtained for wavelengths centered at 550 nm. The experimental determination of the reflectivity of each object used is carried out using KYORITSU make Model 5200 Illuminometer. These experimental results are obtained by maintaining the reflecting surface of each object normal to the optical axes of the emitting LEDs. It can be seen that, for highly reflecting objects, for a given threshold voltage set, the proximity detector gets activated at relatively larger distances. For same threshold voltage chosen, if this object is replaced by other objects with lower and lower reflectivities, then the proximity sensor gets activated at closer and closer proximities i.e. the minimum distance of separation required to cause activation of the proximity output LED will get smaller and smaller. For objects with small reflectivities, these proximity distances are quite small and the sensor reaches its dead band very fast, even for small values of threshold voltages chosen. For example, in our experiments, for object (c), this dead band is reached for a threshold voltage of 0.6 V and for a further increase in this voltage, the system cannot be effectively used for proximity detection. Hence, for effective utilization of this proximity sensor for robot navigation, the objects should be at least having medium or low-medium reflectivites so that the robot can safely avoid them, based on this sensor activation. Our experimentations have also revealed that the sensor system developed can be effectively utilized to detect objects of a minimum dimension of 6 cm × 8 cm or of bigger dimensions.

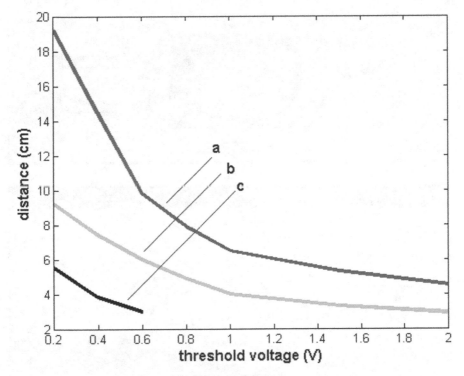

Fig. 4.8. The proximity detector performance curve for objects with (a) reflectivity $p = 33\%$, (b) reflectivity $p = 16\%$, and (c) reflectivity $p = 7.8\%$

4.4 The Intranet-Connectivity for Client-Server Operation

In addition to the two special types of sensor systems, the indigenously developed robot is also equipped with intranet-connectivity where data communication and control command exchange can take place between the laptop mounted on the robot and a remote end PC. In this client-server mode of operation, the robot acts as the server and the remote-end user acts as the client and the communication takes place using Windows based socket programming in TCP/IP protocol.

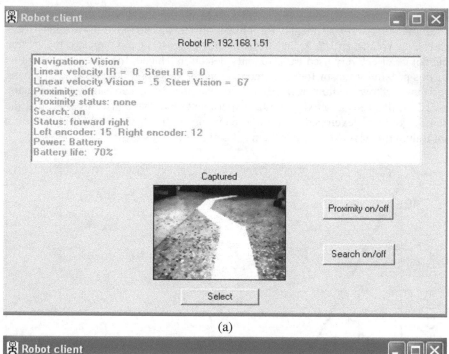

(a)

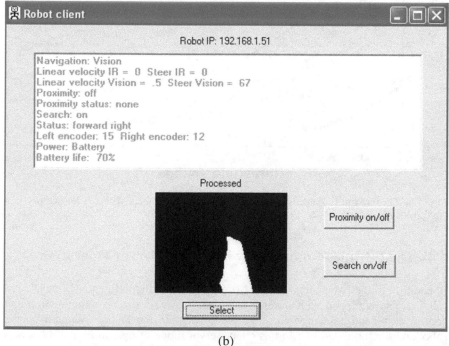

(b)

Fig. 4.9(a) & (b). The GUI-based view from the client-end, at a sample instant, during the robot navigation

The client end can select the robot functionalities like (i) whether the proximity sensors should be used in navigation or not and (ii) whether search mode should be ON or OFF. If the search mode is on, then the stored information for possible steering angle detour is used to guide the robot back to the path, in case the robot leaves the path/line. If the search mode is deselected from the client end, then the robot stops once it leaves the path/line. One can include more such functionalities to add more control flexibilities in remote operation, if it is so desired. The server end can also send both text and image data on receiving "data send_request" from the client. Usually image data is voluminous, and, on receiving a request, the server end first creates an array of all pixel values of an image matrix for transmission. However the entire array of data is not transmitted in a contiguous manner but it is sent in a series of data packets, managed by a low-end device driver. The client end system is also programmed in such a manner that they keep receiving the data packets until a complete image data array is received and then reconstruct the image for display at the client end. The system is developed with an interlocking feature so that the client is not allowed to send a new request, when it is in the process of receiving data packets corresponding to an earlier request. Figure 4.9(a) and Fig. 4.9(b) show an user interface developed in the client end, which show a captured frame and the path/line extracted from this frame, at the server end. As the GUI shows and as mentioned before, the system has the flexibility that, from the client end, one can activate or deactivate the IR proximity sensors, by clicking the button "Proximity on/off". Also one can click the button "Search on/off" which will signify, when the path vanishes from the field-of-view of the robot, whether the robot will continue to take turns in iterative fashion to re-localize itself on the path/line, or will it simply stop further navigation.

4.5 Summary

This chapter described how a low-cost robot can be indigenously developed in the laboratory with special functionalities. The robot system consists of two specially developed microcontroller based sensor systems and also the flexibility of intranet connectivity. Among the two specially developed sensor systems, a PIC microcontroller based IR range finder system is developed where dynamic range enhancement is achieved by adaptively utilizing the IR sensor output to switch one IR LED ON/OFF. This system utilizes an array-based approach to manipulate the burst frequency and duration of IR energy transmission, to enhance accuracy of range finding. Another microcontroller based sensor system designed comprises an optical proximity detection sensor system using white LEDs, an LDR-transistor based electronic circuit and an output LED for Boolean indication of ON/OFF. The scheme is developed using switching mode synchronous detection technique and to facilitate reliable functioning of this circuit under different working conditions, the system is equipped with an external threshold adjustment facility, which an user can advantageously use for fine tuning the performance of the system. Finally the robot is equipped with intranet connectivity for client server operation where the laptop on the robot acts in the slave mode and a remote end PC, in master mode, can command the robot from a remote location for suitable operations.

References

[1]   De Nisi, F., Gonzo, L., Gottardi, M., Stoppa, D., Simoni, A., Angelo-Beraldin, J.: A CMOS sensor optimized for laser spot-position detection. IEEE Sensors Journal 5(6), 1296–1304 (2005)

[2]   Rakshit, A., Chatterjee, A.: A microcontroller based IR range finder system with dynamic range enhancement. IEEE Sensors Journal 10(10), 1635–1636 (2010)

[3]   http://ww1.microchip.com/downloads/en/DeviceDoc/41211D_.pdf

[4]   http://owww.phys.au.dk/elektronik/is1u60.pdf

[5]   Chatterjee, A., Sarkar, G., Rakshit, A.: Neural compensation for a microcontroller based frequency synthesizer-vector voltmeter. IEEE Sensors Journal 11(6), 1427–1428 (2011)

[6]   Chatterjee, A., Sarkar, G., Rakshit, A.: A reinforcement-learning-based fuzzy compensator for a microcontroller-based frequency synthesizer/vector voltmeter. IEEE Transactions on Instrumentation and Measurement 60(9), 3120–3127 (2011)

[7]   Ray, S., Sarkar, G., Chatterjee, A., Rakshit, A.: Development of a microcontroller-based frequency synthesizer cum vector voltmeter. IEEE Sensors Journal 11(4), 1033–1034 (2011)

[8]   Platil, A.: An introduction to synchronous detection, http://measure.feld.cvut.cz/en/system/files/files/en/education/courses/xe38ssd/SynchrDetectBW.pdf

[9]   Borza, D.N.: Mechanical vibration measurement by high-resolution time-averaged digital holography. Measurement Science and Technology 16(9), 1853 (2005), doi:10.1088/0957-0233/16/9/019

[10] Ruggeri, M., Salvatori, G., Rovati, L.: Synchronous phase to voltage converter for true-phase polarimeters. Measurement Science and Technology 16(2), 569 (2005), doi:10.1088/0957-0233/16/2/033

[11] Philp, W.R., Booth, D.J., Shelamoff, A., Linthwaite, M.J.: A simple fibre optic sensor for measurement of vibrational frequencies. Measurement Science and Technology 3(6), 603 (1992), doi:10.1088/0957-0233/3/6/007

[12] PHILIPS Data handbook. Semiconductors and integrated circuits, Part 4b (December 1974)

[13] Min, M., Parve, T.: Improvement of the vector analyser based on two-phase switching mode synchronous detection. Measurement 19(2), 103–111 (1996)

[14] Nirmal Singh, N.: Vision Based Autonomous Navigation of Mobile Robots. Ph.D. Thesis, Jadavpur University, Kolkata, India (2010)

[15] Rakshit, A., Chatterjee, A.: A microcontroller based compensated optical proximity detector employing switching-mode synchronous detection technique. Measurement Science and Technology 23(3) (March 2012), http://m.iopscience.iop.org/0957-0233/23/3/035102

Chapter 5
Sample Implementations of Vision-Based Mobile Robot Algorithms

Abstract. This chapter presents a detailed, step-by-step demonstration of how vision-based navigation modules can be actually implemented in real life, under 32-bit Windows environment. These lessons start with a simple development of capturing image frames from a running video and then gradually proceeds to more complex tasks of incorporating image processing capabilities e.g. filtering techniques, contrast enhancement, adaptive thresholding etc. Then the lessons demonstrate how to extract path for the robot from such images and how a rule-based approach can be utilized to determine left and right wheel speed settings of a differential drive system.

5.1 Introduction

In this chapter Visual Basic based software programming is presented in a step-by-step fashion. Ten lessons are developed for PC based vision-based navigation programming. Low-cost webcam is used for capturing streaming video.

Visual Basic version 6 (VB6) [1-2] is used for windows based programming.

The first lesson 'Lesson 1' demonstrates how to capture image frames from streaming video from a low-cost webcam and examine pixel (picture element) values with the help of mouse pointer. RGB (Red-Green-Blue) to gray-scale conversion is also done in a pixel-by-pixel manner. A 'Format' menu is provided for selecting the image frame size to 160x120. Windows 32-bit API (Application Programming Interface) calls [3] are adopted for faster processing.

The second lesson 'Lesson 2' demonstrates how to process captured image frames from streaming video. Options are provided for RGB to gray-scale conversion and subsequent low-pass filtering [4].

The third lesson 'Lesson 3' shows the method of contrast enhancement by histogram stretching technique [4] under poor lighting conditions.

The fourth lesson 'Lesson 4' introduces geometric-mean filter [4] to smooth and suppress image detail to simplify the extraction of required white path for navigation.

The fifth lesson 'Lesson 5' applies an adaptive threshold operation to extract white path under varying illumination conditions. A selectable reference pixel determines the centre of path to be extracted.

A. Chatterjee et al.: Vision Based Autonomous Robot Navigation, SCI 455, pp. 101–142.
springerlink.com   © Springer-Verlag Berlin Heidelberg 2013

The sixth lesson 'Lesson 6' introduces a cleaning operation to remove unwanted objects detected during threshold operation.

The next lesson 'Lesson 7' introduces an option for selection of path color white or black. For black path color option, the gray-scale image frame is first converted to negative image, so that black objects become white and then processed as usual as discussed in 'Lesson 6'.

The eighth lesson 'Lesson 8' is targeted for white or black path finding for navigation with a fixed reference pixel.

The next lesson 'Lesson 9' introduces a rule-based approach to determine left and right wheel speed settings of a differential drive system for navigation. Pictorial representation of navigation direction is done with appropriate image file.

Finally in the last lesson 'Lesson 10' sound output is added to draw attention during navigation.

Source codes are available for Visual Basic version 6 and Visual Basic dot net version 2010 compiler from 'http://extras.springer.com'.

Executable codes are also provided for testing the performance of programs when compilers are not available with the reader. Only run-time executables are needed which are freely available from Microsoft.

5.2 Lesson 1

Objective: To develop a VB6 program to capture webcam streaming video.

Following steps summarize the program development.

1. All necessary Application Programming Interface (API) calls are declared in 'Webcam1.bas' module. It is necessary to include this module in 'Form1' of the VB6 program.
2. AVICAP32.DLL is used to capture webcam streaming video through proper API call. The webcam video format should be either RGB24 or YUY2.
3. Under Form1 two 'Picture Box' controls are added, 'Picture1' to preview streaming video at 30 frames per second and 'Picture2' to capture image from streaming video as clipboard data at a regular interval of 10mS with the help of 'Timer1' control.
4. Two command buttons, namely, 'Capture' and 'Close' are added under 'Form1' to control image capturing process. The command button names are 'cmdCapture' and 'cmdClose' respectively.
5. A menu item 'Format' is added in 'Form1' to set the image size to 160x120 pixels.
6. Any captured pixel may be examined with the mouse pointer over 'picture2' image. The mouse cursor is changed to 'cross' to facilitate pixel examination.
7. Pixel color is obtained through the 'GetPixel' API call.
8. Red (R), Green (G) and Blue (B) vales are obtained from 'Color' by calling three functions 'GetRed', 'GetGreen' and 'GetBlue' functions as follows: GetRed = Color And 255, GetGreen = (Color And 65280) \ 256 and GetBlue = (Color And 16711680) \ 65535.
9. Three text boxes, namely, 'Text1', 'Text2' and 'Text3' are added to examine 8-bit Red (R), Green (G) and Blue (B) values of the selected pixel.

10. Two text boxes, namely, 'Text4' and 'Text5', are incorporated to monitor 'X' and 'Y' coordinates of the selected pixel.
11. A text box 'Text6' is added to view 8-bit gray value of the selected pixel from its RGB values according to the formula: gray = 0.2125 * red + 0.7154 * green + 0.0721 * blue.
12. A second timer 'Timer2' control is added to remove textbox data within 10mS when the mouse pointer is not positioned over 'Picture2' picture box.

Following text shows the listing of 'Webcam1.bas' module.

```
Global Const WS_CHILD As Long = &H40000000
Global Const WS_VISIBLE As Long = &H10000000
Global Const WM_USER = 1024
Global Const WM_CAP_DRIVER_CONNECT = WM_USER + 10
Global Const WM_CAP_SET_PREVIEW = WM_USER + 50
Global Const WM_CAP_SET_PREVIEWRATE = WM_USER + 52
Global Const WM_CAP_DRIVER_DISCONNECT As Long = WM_USER + 11
Global Const WM_CAP_DLG_VIDEOFORMAT As Long = WM_USER + 41
Global Const WM_CAP_GET_FRAME As Long = 1084
Global Const WM_CAP_COPY As Long = 1054
Global Const WM_CAP_SET_SCALE As Integer = WM_USER + 53
Global Const SWP_NOMOVE As Integer = 2
Global Const SWP_NOZORDER As Integer = 4
Global Const HWND_BOTTOM As Integer = 1


Declare Function SendMessage Lib "user32" Alias "SendMessageA" _
    (ByVal hwnd As Long, ByVal wMsg As Long, ByVal wParam As _
    Long, ByVal lParam As Long) As Long
Declare Function capCreateCaptureWindow Lib "avicap32.dll" Alias _
    "capCreateCaptureWindowA" (ByVal a As String, ByVal b As Long, _
    ByVal c As Integer, ByVal d As Integer, ByVal e As Integer, _
    ByVal f As Integer, ByVal g As Long, ByVal h As Integer) As Long
Declare Function SetWindowPos Lib "user32" (ByVal hwnd As Long, _
    ByVal hWndInsertAfter As Long, ByVal x As Long, ByVal y As Long, _
    ByVal cx As Long, ByVal cy As Long, ByVal wFlags As Long) As Long
Declare Function GetPixel Lib "gdi32" (ByVal hdc As Long, _
    ByVal x As Long, ByVal y As Long) As Long
```

Following figure shows the 'Form1' layout.

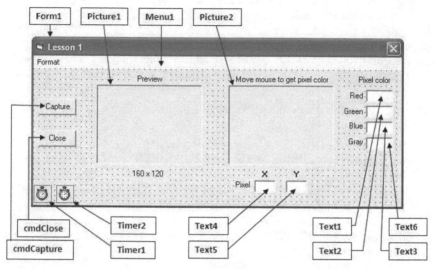

Following text shows the listing of 'Form1' code.

```
Dim hwdc As Long
Dim startcap As Boolean
Dim mflag As Boolean

Private Sub cmdCapture_Click()
   hwdc = capCreateCaptureWindow("Webcam Vision System", WS_CHILD _
Or WS_VISIBLE, 0, 0, 160, 120, Picture1.hwnd, 0)
   If (hwdc <> 0) Then
      Clipboard.Clear
      If SendMessage(hwdc, WM_CAP_DRIVER_CONNECT, 0, 0) Then
         SendMessage hwdc, WM_CAP_SET_SCALE, True, 0
         SendMessage hwdc, WM_CAP_SET_PREVIEWRATE, 30, 0
         SendMessage hwdc, WM_CAP_SET_PREVIEW, 1, 0
         SetWindowPos hwdc, HWND_BOTTOM, 0, 0, 160, 120, SWP_NOMOVE _
         Or SWP_NOZORDER
         startcap = True
         cmdCapture.Enabled = False
         cmdClose.Enabled = True
         Timer1.Enabled = True
         Menu1.Enabled = True
         Picture2.Visible = True
         Label1.Visible = True
         Label2.Visible = True
         Label3.Visible = True
         Label4.Visible = True
         Label5.Visible = True
         Label6.Visible = True
```

```
                    Label7.Visible = True
                    Label9.Visible = True
                    Label11.Visible = True
                    Text1.Visible = True
                    Text2.Visible = True
                    Text3.Visible = True
                    Text4.Visible = True
                    Text5.Visible = True
                    Text6.Visible = True
            Else
                    MsgBox ("No Webcam found!")
                    startcap = False
            End If
        End If
End Sub

Private Sub cmdClose_Click()
    If startcap = True Then
        SendMessage hwdc, WM_CAP_DRIVER_DISCONNECT, 0, 0
        startcap = False
        cmdCapture.Enabled = True
        cmdClose.Enabled = False
        Timer1.Enabled = False
        Menu1.Enabled = False
        Picture2.Visible = False
        Label1.Visible = False
        Label2.Visible = False
        Label3.Visible = False
        Label4.Visible = False
        Label5.Visible = False
        Label6.Visible = False
        Label7.Visible = False
        Label9.Visible = False
        Label11.Visible = False
        Text1.Visible = False
        Text2.Visible = False
        Text3.Visible = False
        Text4.Visible = False
        Text5.Visible = False
        Text6.Visible = False
    End If
End Sub
```

```
Private Sub Form_Load()
    If App.PrevInstance = True Then End     ' multiple instances are not allowed
    cmdCapture.Enabled = True
    cmdClose.Enabled = False
    Picture1.AutoSize = True
    Picture2.AutoSize = True
    Timer1.Interval = 10
    Timer2.Interval = 10
    Menu1.Enabled = False
    mflag = False
    Picture2.Visible = False
    Picture2.MousePointer = 2                      ' cross cursor
    Label1.Visible = False
    Label2.Visible = False
    Label3.Visible = False
    Label4.Visible = False
    Label5.Visible = False
    Label6.Visible = False
    Label7.Visible = False
    Label9.Visible = False
    Label11.Visible = False
    Text1.Visible = False
    Text2.Visible = False
    Text3.Visible = False
    Text4.Visible = False
    Text5.Visible = False
    Text6.Visible = False
End Sub

Private Function GetRed(ByVal Color As Long)
    GetRed = Color And 255
End Function

Private Function GetGreen(ByVal Color As Long)
    GetGreen = (Color And 65280) \ 256
End Function

Private Function GetBlue(ByVal Color As Long)
    GetBlue = (Color And 16711680) \ 65535
End Function

Private Sub Form_MouseMove(Button As Integer, Shift As Integer, _
    x As Single, y As Single)
    mflag = False                    ' mouse pointer in form but not in picture box
End Sub
Private Sub Menu1_Click()
```

```
    If startcap = True Then
        SendMessage hwdc, WM_CAP_DLG_VIDEOFORMAT, 0, 0
    End If
End Sub

Private Sub Picture2_MouseMove(Button As Integer, Shift As Integer, _
    x As Single, y As Single)
    Dim Color As Long
    Dim red As Byte
    Dim blue As Byte
    Dim green As Byte
    Dim gray As Byte
    Dim xp As Long
    Dim yp As Long

    xp = x / Screen.TwipsPerPixelX
    yp = y / Screen.TwipsPerPixelY
    Color = GetPixel(Picture2.hdc, xp, yp)
    red = GetRed(Color)
    green = GetGreen(Color)
    blue = GetBlue(Color)
    gray = 0.2125 * red + 0.7154 * green + 0.0721 * blue
    Text1.Text = red
    Text2.Text = green
    Text3.Text = blue
    Text4.Text = xp
    Text5.Text = yp
    Text6.Text = gray
    mflag = True                              ' mouse pointer in picture box
End Sub

Private Sub Timer1_Timer()
    SendMessage hwdc, WM_CAP_GET_FRAME, 0, 0
    SendMessage hwdc, WM_CAP_COPY, 0, 0
    Picture2.Picture = Clipboard.GetData
    SendMessage hwdc, WM_CAP_SET_PREVIEW, 1, 0
End Sub

Private Sub Timer2_Timer()
    If mflag = False Then                     ' no mouse pointer in picture box
        Text1.Text = ""
        Text2.Text = ""
        Text3.Text = ""
        Text4.Text = ""
```

```
        Text5.Text = ""
        Text6.Text = ""
    End If
End Sub
```

To execute the program the capture button has to be pressed. If any webcam is available then preview is available in picture box 'Picture1'. If the size of the captured image does not fit in the picture box 'Picture2' then the image size has to be changed to 160x120 by activating the 'Format' menu.

If no webcam is available then a message box will appear with a message "No webcam found!"

5.3 Lesson 2

Objective: To develop a VB6 program to capture and process webcam streaming video for conversion to gray scale image and subsequent low-pass image filtering.

Following steps summarize the program development.

1. All necessary API calls are declared in 'Webcam2.bas' module. It is necessary to include this module in 'Form1' of the VB6 program.
2. AVICAP32.DLL is used to capture webcam streaming video through proper API call. The webcam video format should be either RGB24 or YUY2.
3. Under Form1 two 'Picture Box' controls are added, 'Picture1' to capture image as clipboard data from streaming video at a regular interval of 10mS and 'Picture2' to process image from captured image at the same rate with the help of 'Timer1' control.
4. A menu item 'Format' is added in 'Form1' to set the image size to 160x120 pixels.
5. From 'Picture1' image pixel data information is obtained through 'GetObject' API call.
6. Pixel array 'Pbytes(c, x, y)', an 8-bit array, is obtained through 'GetBitmapBits' API call under 'Timer1' control. Each element of 'Pbytes' contains 8-bit RGB color information of each pixel at 'x' and 'y' image co-ordinate. 'c' stands for color; c:2 for red, c:1 for green and c:0 for blue.
7. Pixel array is processed according to option controls 'Option1' or 'Option2'.
8. If 'Option1' is selected then pixel array is processed as gray scale image with the help of procedure 'Gray' and displayed in picture box 'Picture2' through 'SetBitmapBits' API call.
9. If 'Option2' is selected then pixel array is processed first to gray scale image as in step 8 and then low-pass filtered with the help of procedure 'Lowpass' and then displayed in 'Picture2'.

Option1   Option2

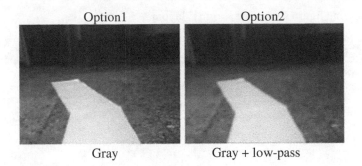

Gray   Gray + low-pass

Following text shows the listing of 'Webcam2.bas' module.

```
Global Const WS_CHILD As Long = &H40000000
Global Const WS_VISIBLE As Long = &H10000000
Global Const WM_USER = 1024
Global Const WM_CAP_DRIVER_CONNECT = WM_USER + 10
Global Const WM_CAP_SET_PREVIEW = WM_USER + 50
Global Const WM_CAP_SET_PREVIEWRATE = WM_USER + 52
Global Const WM_CAP_DRIVER_DISCONNECT As Long = WM_USER + 11
Global Const WM_CAP_DLG_VIDEOFORMAT As Long = WM_USER + 41
Global Const WM_CAP_GET_FRAME As Long = 1084
Global Const WM_CAP_COPY As Long = 1054
Global Const WM_CAP_SET_SCALE As Integer = WM_USER + 53
Global Const SWP_NOMOVE As Integer = 2
Global Const SWP_NOZORDER As Integer = 4
Global Const HWND_BOTTOM As Integer = 1


Declare Function SendMessage Lib "user32" Alias "SendMessageA" (ByVal hwnd _
    As Long, ByVal wMsg As Long, ByVal wParam As Long, ByVal lParam As_
    Long) As Long Declare Function capCreateCaptureWindow Lib _
    "avicap32.dll" Alias "capCreateCaptureWindowA" (ByVal nWindowName _
    As String, ByVal nStyle As Long, ByVal nx As Integer, ByVal ny As Integer, _
    ByVal nWidth As Integer, ByVal nHeight As Integer, ByVal nHwnd As Long, _
    ByVal nId As Integer) As Long
Declare Function SetWindowPos Lib "user32" (ByVal hwnd As Long, _
    ByVal hWndInsertAfter As Long, ByVal x As Long, ByVal y As Long, _
    ByVal cx As Long, ByVal cy As Long, ByVal wFlags As Long) As Long
Declare Function GetObject Lib "gdi32" Alias "GetObjectA" (ByVal hObject _
    As Long, ByVal nCount As Long, lpObject As Any) As Long
Declare Function GetBitmapBits Lib "gdi32" (ByVal hBitmap As Long, _
    ByVal dwCount As Long, lpBits As Any) As Long
Declare Function SetBitmapBits Lib "gdi32" (ByVal hBitmap As Long, _
    ByVal dwCount As Long, lpBits As Any) As Long
```

Following figure shows the 'Form1' layout.

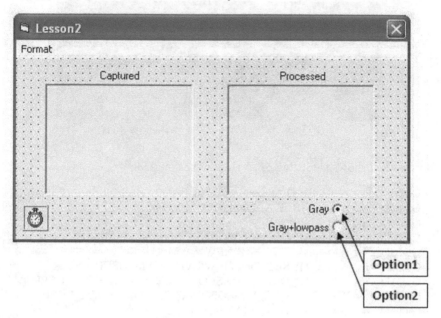

Following text shows the listing of 'Form1' code.

```
Dim hwdc As Long
Dim startcap As Boolean
Private Type Bitmap
   bmType As Long
   bmWidth As Long
   bmHeight As Long
   bmWidthBytes As Long
   bmPlanes As Integer
   bmBitsPixel As Integer
   bmBits As Long
End Type
Dim Pbytes() As Byte, Pinfo As Bitmap
Dim x As Long, y As Long

Private Sub Form_Load()
   If App.PrevInstance = True Then End
   Picture1.AutoSize = True
   Picture2.AutoSize = True
   Picture1.ScaleMode = vbPixels
   Picture2.ScaleMode = vbPixels
   Timer1.Interval = 10

 hwdc = capCreateCaptureWindow("Webcam Vision System", WS_CHILD _
Or WS_VISIBLE, 0, 0, 160, 120, Picture1.hwnd, 0)
```

```
    If (hwdc <> 0) Then
      Clipboard.Clear
      If SendMessage(hwdc, WM_CAP_DRIVER_CONNECT, 0, 0) Then
        SendMessage hwdc, WM_CAP_SET_SCALE, 1, 0
        SendMessage hwdc, WM_CAP_SET_PREVIEWRATE, 30, 0
        SendMessage hwdc, WM_CAP_SET_PREVIEW, 1, 0
        SetWindowPos hwdc, HWND_BOTTOM, 0, 0, 160, 120, _
        SWP_NOMOVE Or SWP_NOZORDER
        SendMessage hwdc, WM_CAP_GET_FRAME, 0, 0
        SendMessage hwdc, WM_CAP_COPY, 0, 0
        Picture1.Picture = Clipboard.GetData
        GetObject Picture1.Picture, Len(Pinfo), Pinfo
        ReDim Pbytes(0 To (Pinfo.bmBitsPixel \ 8) - 1, 0 To Pinfo.bmWidth - 1, _
        0 To Pinfo.bmHeight - 1)
        Picture2.height = Picture1.height
        Picture2.width = Picture1.width
        Timer1.Enabled = True
        startcap = True
      Else
        MsgBox "No Webcam found!", OK, ""
        startcap = False
        Unload Me
      End If
    Else
      Unload Me
    End If
End Sub

Private Sub Gray(width As Long, height As Long)
    Dim G As Byte
    For x = 0 To width - 1
      For y = 0 To height - 1
        G = 0.2125 * CDbl(Pbytes(2, x, y)) + 0.7154 * CDbl(Pbytes(1, x, y)) + _
            0.0721 * CDbl(Pbytes(0, x, y))
        Pbytes(2, x, y) = G              'Red
        Pbytes(1, x, y) = G              'Green
        Pbytes(0, x, y) = G              'Blue
      Next y
    Next x
End Sub

Private Sub Lowpass(width As Long, height As Long)
    Dim R As Long
    Dim c, d, e, f As Long
    For x = 0 To width - 1
      For y = 0 To height - 1
```

```
            c = x - 1
            d = x + 1
            e = y - 1
            f = y + 1
            If c < 0 Then c = width - 1
            If d = width Then d = 0
            If e < 0 Then e = height - 1
            If f = height Then f = 0
            R = Pbytes(2, x, e)
            R = R + CLng(Pbytes(2, c, y))
            R = R + 2 * CLng(Pbytes(2, x, y))
            R = R + CLng(Pbytes(2, d, y))
            R = R + CLng(Pbytes(2, x, f))
            R = R / 6                          '3x3 low pass
            Pbytes(2, x, y) = R
            Pbytes(1, x, y) = R
            Pbytes(0, x, y) = R
        Next y
     Next x
End Sub

Private Sub Form_Terminate()
     If startcap = True Then
        SendMessage hwdc, WM_CAP_DRIVER_DISCONNECT, 0, 0
        startcap = False
        Timer1.Enabled = False
     End If
End Sub

Private Sub Form_Unload(Cancel As Integer)
     If startcap = True Then
        SendMessage hwdc, WM_CAP_DRIVER_DISCONNECT, 0, 0
        startcap = False
        Timer1.Enabled = False
     End If
End Sub

Private Sub Menu_Click()
     If startcap = True Then
        SendMessage hwdc, WM_CAP_DLG_VIDEOFORMAT, 0, 0
     End If
End Sub

Private Sub Timer1_Timer()
     Timer1.Enabled = False
     SendMessage hwdc, WM_CAP_GET_FRAME, 0, 0
     SendMessage hwdc, WM_CAP_COPY, 0, 0
```

```
        Picture1.Picture = Clipboard.GetData
        GetBitmapBits Picture1.Picture, Pinfo.bmWidthBytes * Pinfo.bmHeight, _
            Pbytes(0, 0, 0)
        If Option1.Value = True Then Gray Picture1.ScaleWidth, Picture1.ScaleHeight
        If Option2.Value = True Then
            Gray Picture1.ScaleWidth, Picture1.ScaleHeight
            Lowpass Picture1.ScaleWidth, Picture1.ScaleHeight
        End If
        SetBitmapBits Picture2.Image, Pinfo.bmWidthBytes * Pinfo.bmHeight, _
            Pbytes(0, 0, 0)
        Picture2.Refresh
        Picture2.Picture = Picture2.Image
        Timer1.Enabled = True
End Sub
```

Low-pass filtering is performed with a 2-D FIR filer mask of size 3x3 as stated below:

$$\frac{1}{6} \begin{bmatrix} 0 & 1 & 0 \\ 1 & 2 & 1 \\ 0 & 1 & 0 \end{bmatrix}$$

Circular 2-D convolution is performed with the above mask to preserve the image size before and after filtering with minimum amount of distortion.

If the size of the captured image does not fit in the picture box then the image size has to be changed to 160x120 by activating the 'Format' menu. If no webcam is available then a message box will appear with a message "No webcam found!"

5.4 Lesson 3

Objective: To develop a VB6 program to capture and process webcam streaming video for conversion to gray scale image, low-pass image filtering and contrast enhancement.

Following steps summarize the program development.

1. All necessary API calls are declared in 'Webcam3.bas' module, same as 'Webcam2.bas', as mentioned in Lesson 2. It is necessary to include this module in 'Form1' of the VB6 program.
2. AVICAP32.DLL is used to capture webcam streaming video through proper API call. The webcam video format should be either RGB24 or YUY2.
3. Under Form1 two 'Picture Box' controls are added, 'Picture1' to capture image as clipboard data from streaming video at a regular interval of 10mS and 'Picture2' to process image from captured image at the same rate with the help of 'Timer1' control.
4. A menu item 'Format' is added in 'Form1' to set the image size to 160x120 pixels.

5. From 'Picture1' image pixel data information is obtained through 'GetObject' API call.
6. Pixel array 'Pbytes(c, x, y)', an 8-bit array, is obtained through 'GetBitmapBits' API call under 'Timer1' control. Each element of 'Pbytes' contains 8-bit RGB color information of each pixel at 'x' and 'y' image co-ordinate. 'c' stands for color; c:2 for red, c:1 for green and c:0 for blue.
7. Pixel array is processed according to option controls 'Option1', 'Option2' or 'Option3'.
8. If 'Option1' is selected then pixel array is processed as gray scale image with the help of procedure 'Gray' and displayed in picture box 'Picture2' through 'SetBitmapBits' API call.
9. If 'Option2' is selected then pixel array is processed first to gray scale image as in step 8 and then low-pass filtered with the help of procedure 'Lowpass' and then displayed in 'Picture2'.
10. If 'Option3' is selected then array is low-pass filtered as in step 9 and then processed for contrast enhancement using histogram stretching technique with the help of procedure 'Contrast' and then displayed in 'Picture2'.

Option1   Option2   Option3

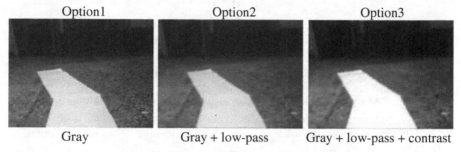

Gray   Gray + low-pass   Gray + low-pass + contrast

Following figure shows the 'Form1' layout.

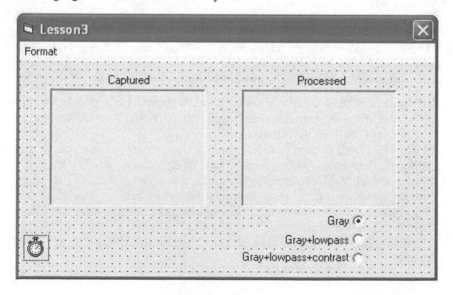

Following text shows the listing of 'Contrast' and 'Timer1' procedure code. For rest of the code refer to Lesson 2.

```
Private Sub Contrast(width As Long, height As Long)
    Dim R As Long                          'histogram stretching
    Dim pmax, pmin As Long
    pmax = 0
    pmin = 255
    For x = 0 To width - 1
      For y = 0 To height - 1
        If pmax <= CLng(Pbytes(2, x, y)) Then pmax = Pbytes(2, x, y)
        If pmin >= CLng(Pbytes(2, x, y)) Then pmin = Pbytes(2, x, y)
      Next y
    Next x
    For x = 0 To width - 1
      For y = 0 To height - 1
        R = Pbytes(2, x, y)
        If pmax > pmin Then R = (((R - pmin) * 255) / (pmax - pmin)) + pmin / 4
        If R < 0 Then R = 0
        If R > 255 Then R = 255
        Pbytes(2, x, y) = R
        Pbytes(1, x, y) = R
        Pbytes(0, x, y) = R
      Next y
    Next x
End Sub


Private Sub Timer1_Timer()
    Timer1.Enabled = False
    SendMessage hwdc, WM_CAP_GET_FRAME, 0, 0
    SendMessage hwdc, WM_CAP_COPY, 0, 0
    Picture1.Picture = Clipboard.GetData
    GetBitmapBits Picture1.Picture, Pinfo.bmWidthBytes * Pinfo.bmHeight, _
        Pbytes(0, 0, 0)
    If Option1.Value = True Then Gray Picture1.ScaleWidth, Picture1.ScaleHeight
    If Option2.Value = True Then
      Gray Picture1.ScaleWidth, Picture1.ScaleHeight
      Lowpass Picture1.ScaleWidth, Picture1.ScaleHeight
    End If
    If Option3.Value = True Then
      Gray Picture1.ScaleWidth, Picture1.ScaleHeight
      Lowpass Picture1.ScaleWidth, Picture1.ScaleHeight
      Contrast Picture1.ScaleWidth, Picture1.ScaleHeight
    End If
```

```
        SetBitmapBits Picture2.Image, Pinfo.bmWidthBytes * Pinfo.bmHeight, _
            Pbytes(0, 0, 0)
        Picture2.Refresh
        Picture2.Picture = Picture2.Image
        Timer1.Enabled = True
    End Sub
```

If the size of the captured image does not fit in the picture box then the image size has to be changed to 160x120 by activating the 'Format' menu. If no webcam is available then a message box will appear with a message "No webcam found!"

5.5 Lesson 4

Objective: To develop a VB6 program to capture and process webcam streaming video for conversion to gray scale image, low-pass filtering, contrast enhancement and geometric-mean filtering.

Following steps summarize the program development.

1. All necessary API calls are declared in 'Webcam4.bas' module, same as 'Webcam3.bas', as mentioned in Lesson 3. It is necessary to include this module in 'Form1' of the VB6 program.
2. AVICAP32.DLL is used to capture webcam streaming video through proper API call. The webcam video format should be either RGB24 or YUY2.
3. Under Form1 two 'Picture Box' controls are added, 'Picture1' to capture image as clipboard data from streaming video at a regular interval of 10mS and 'Picture2' to process image from captured image at the same rate with the help of 'Timer1' control.
4. A menu item 'Format' is added in 'Form1' to set the image size to 160x120 pixels.
5. From 'Picture1' image pixel data information is obtained through 'GetObject' API call.
6. Pixel array 'Pbytes(c, x, y)', an 8-bit array, is obtained through 'GetBitmapBits' API call under 'Timer1' control. Each element of 'Pbytes' contains 8-bit RGB color information of each pixel at 'x' and 'y' image co-ordinate. 'c' stands for color; c:2 for red, c:1 for green and c:0 for blue.
7. Pixel array is processed according to option controls 'Option1', 'Option2', 'Option3' or 'Option4'.
8. If 'Option1' is selected then pixel array is processed as gray scale image with the help of procedure 'Gray' and displayed in picture box 'Picture2' through 'SetBitmapBits' API call.
9. If 'Option2' is selected then pixel array is processed first to gray scale image as in step 8 and then low-pass filtered with the help of procedure 'Lowpass' and then displayed in 'Picture2'.

10. If 'Option3' is selected then array is low-pass filtered as in step 9 and then processed for contrast enhancement using histogram stretching technique with the help of procedure 'Contrast' and then displayed in 'Picture2'.

11. If 'Option4' is selected then array is processed for contrast enhancement as in step 10 and then processed for geometric-mean filtering with the help of procedure 'Geometricmean' and then displayed in 'Picture2'. Options are provided for increasing the number of cascaded Geometric-mean filters and the size of mask for each filter.

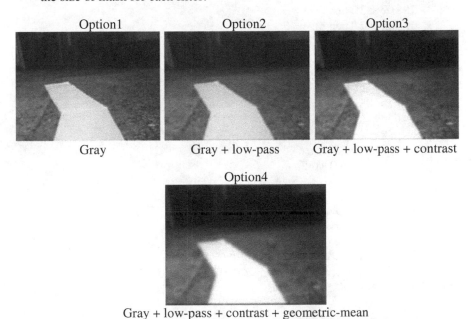

Option1	Option2	Option3
Gray	Gray + low-pass	Gray + low-pass + contrast

Option4

Gray + low-pass + contrast + geometric-mean

Following figure shows the 'Form1' layout.

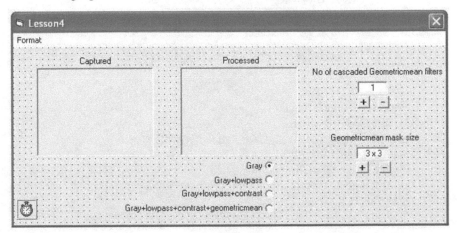

Following text shows the listing of 'Geometricmean' and 'Timer1' procedure code. For rest of the code refer to Lesson 3.

```
Private Sub Geometricmean(width As Long, height As Long, Size As Long)
    Dim R, S As Long
    Dim i, j As Long
    Dim c, d As Long
    Dim w1, h1 As Long
    If Size < 3 Then Size = 3
    If Size > 7 Then Size = 7
    If (Size And 1) = 0 Then Size = Size + 1       'even to odd conversion
    S = Size * Size
    w1 = width - 1
    h1 = height - 1
    For x = 0 To w1
      For y = 0 To h1
        R = 1
        For i = 0 To Size - 1
          For j = 0 To Size - 1
            c = x + i - ((Size - 1) / 2)
            If c < 0 Then c = width + c
            If c > w1 Then c = c - w1
            d = y + j - ((Size - 1) / 2)
            If d < 0 Then d = height + d
            If d > h1 Then d = d - h1
            R = R * CLng(Pbytes(2, c, d))
          Next j
        Next i
        R = R ^ (1# / S)
        If R > 255 Then R = 255
        Pbytes(2, x, y) = R
        Pbytes(1, x, y) = R
        Pbytes(0, x, y) = R
      Next y
    Next x
End Sub

Private Sub Timer1_Timer()
  Timer1.Enabled = False
  SendMessage hwdc, WM_CAP_GET_FRAME, 0, 0
  SendMessage hwdc, WM_CAP_COPY, 0, 0
  Picture1.Picture = Clipboard.GetData
  GetBitmapBits Picture1.Picture, Pinfo.bmWidthBytes * Pinfo.bmHeight, _
    Pbytes(0, 0, 0)
  If Option1.Value = True Then Gray Picture1.ScaleWidth, Picture1.ScaleHeight
```

```
        If Option2.Value = True Then
            Gray Picture1.ScaleWidth, Picture1.ScaleHeight
            Lowpass Picture1.ScaleWidth, Picture1.ScaleHeight
        End If
        If Option3.Value = True Then
            Gray Picture1.ScaleWidth, Picture1.ScaleHeight
            Lowpass Picture1.ScaleWidth, Picture1.ScaleHeight
            Contrast Picture1.ScaleWidth, Picture1.ScaleHeight
        End If
        If Option4.Value = True Then
            Gray Picture1.ScaleWidth, Picture1.ScaleHeight
            Lowpass Picture1.ScaleWidth, Picture1.ScaleHeight
            Contrast Picture1.ScaleWidth, Picture1.ScaleHeight
            For i = 1 To Val(Text4.Text)
                Geometricmean Picture1.ScaleWidth, Picture1.ScaleHeight, gms
            Next i
        End If
        SetBitmapBits Picture2.Image, Pinfo.bmWidthBytes * Pinfo.bmHeight, _
            Pbytes(0, 0, 0)
        Picture2.Refresh
        Picture2.Picture = Picture2.Image
        Timer1.Enabled – True
    End Sub
```

If the size of the captured image does not fit in the picture box then the image size has to be changed to 160x120 by activating the 'Format' menu. If no webcam is available then a message box will appear with a message "No webcam found!"

5.6 Lesson 5

Objective: To develop a VB6 program to capture and process webcam streaming video for conversion to gray scale image, low-pass filtering, contrast enhancement, geometric-mean filtering and an adaptive threshold operation to extract white path from the captured image under varying illumination conditions.

Following steps summarize the program development.

1. All necessary API calls are declared in 'Webcam5.bas' module, same as 'Webcam4.bas', as mentioned in Lesson 4. It is necessary to include this module in 'Form1' of the VB6 program.
2. AVICAP32.DLL is used to capture webcam streaming video through proper API call. The webcam video format should be either RGB24 or YUY2.
3. Under Form1 two 'Picture Box' controls are added, 'Picture1' to capture image as clipboard data from streaming video at a regular interval of 10mS and 'Picture2' to process image from captured image at the same rate with the help of 'Timer1' control.

4. A menu item 'Format' is added in 'Form1' to set the image size to 160x120 pixels.
5. From 'Picture1' image pixel data information is obtained through 'GetObject' API call.
6. Pixel array 'Pbytes(c, x, y)', an 8-bit array, is obtained through 'GetBitmapBits' API call under 'Timer1' control. Each element of 'Pbytes' contains 8-bit RGB color information of each pixel at 'x' and 'y' image co-ordinate. 'c' stands for color; c:2 for red, c:1 for green and c:0 for blue.
7. Pixel array is processed according to option controls 'Option1', 'Option2, 'Option3', 'Option4' or 'Option5'.
8. If 'Option1' is selected then pixel array is processed as gray scale image with the help of procedure 'Gray' and displayed in picture box 'Picture2' through 'SetBitmapBits' API call.
9. If 'Option2' is selected then pixel array is processed first to gray scale image as in step 8 and then low-pass filtered with the help of procedure 'Lowpass' and then displayed in 'Picture2'.
10. If 'Option3' is selected then array is low-pass filtered as in step 9 and then processed for contrast enhancement using histogram stretching technique with the help of procedure 'Contrast' and then displayed in 'Picture2'.
11. If 'Option4' is selected then array is processed for contrast enhancement as in step 10 and then processed for geometric-mean filtering with the help of procedure 'Geometricmean' and then displayed in 'Picture2'. Options are provided for increasing the number of cascaded Geometric-mean filters and the size of mask for each filter.
12. If 'Option5' is selected then an adaptive threshold operation is performed with the help of the procedure 'Adaptive Threshold' and then displayed in 'Picture2'. First the white line width around a reference pixel [at the nominal position (80,110)] is determined with the procedure 'WhiteLineWidth'. If both left and right path width around the reference pixel are found be less than 'MIN_PATH_WIDTH' value then a parameter 'delta' is adjusted to increase the path width by decreasing the threshold value within a range 'delta_max'. Then the procedure 'Threshold' computes new image and the above sequence of operations repeats until a valid white path is obtained.

Option5

Gray + low-pass + contrast + geometric-mean + threshold

Following figure shows the 'Form1' layout.

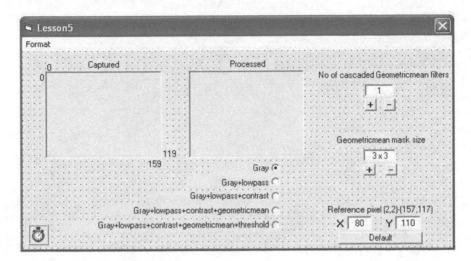

Following text shows the listing of 'AdaptiveThreshold', 'WhiteLineWidth', 'Threshold' and 'Timer1' procedure code. For rest of the code refer to Lesson 4.

```
Private Sub AdaptiveThreshold(width As Long, xr As Long, yr As Long)
    Dim i As Integer
    WhiteLineWidth width, xr, yr
    If PixelCountLeft < MIN_PATH_WIDTH And PixelCountRight < _
    MIN_PATH_WIDTH Then
      delta = delta + 0.2
      If delta > delta_max Then
        delta = delta_max
      Else
        GoTo atc
      End If
      If delta < 1# Then delta = 1#
    End If
    delta = delta - 0.5
atc:
    i = Pbytes(2, xr, yr)
    If i > (255 - (2 * delta)) Then
      If i > (255 - delta) Then i = (255 - delta)
      Threshold Picture1.ScaleWidth, Picture1.ScaleHeight, i - CInt(delta), _
        i + CInt(delta)
    Else
      Threshold Picture1.ScaleWidth, Picture1.ScaleHeight, 255, 255
    End If
End Sub
```

```
Private Sub WhiteLineWidth(width As Long, xr As Long, yr As Long)
    Dim pcl1, pcl2, pcl3, pcr1, pcr2, pcr3 As Integer
    PixelCountLeft = 0: PixelCountRight = 0
    y = yr
    pcl1 = 0: pcr1 = 0
    For x = xr To 0 Step -1
        If Pbytes(2, x, y) > 250 Then
            pcl1 = pcl1 + 1
        End If
    Next x
    For x = (xr + 1) To (width - 1)
        If Pbytes(2, x, y) > 250 Then
            pcr1 = pcr1 + 1
        End If
    Next x
    y = yr - 1
    pcl2 = 0: pcr2 = 0
    For x = xr To 0 Step -1
        If Pbytes(2, x, y) > 250 Then
            pcl2 = pcl2 + 1
        End If
    Next x
    For x = (xr + 1) To (width - 1)
        If Pbytes(2, x, y) > 250 Then
            pcr2 = pcr2 + 1
        End If
    Next x
    y = yr + 1
    pcl3 = 0: pcr3 = 0
    For x = xr To 0 Step -1
        If Pbytes(2, x, y) > 250 Then
            pcl3 = pcl3 + 1
        End If
    Next x
    For x = (xr + 1) To (width - 1)
        If Pbytes(2, x, y) > 250 Then
            pcr3 = pcr3 + 1
        End If
    Next x

    PixelCountLeft = (pcl1 + pcl2 + pcl3) / 3
    PixelCountRight = (pcr1 + pcr2 + pcr3) / 3
End Sub
```

```
Private Sub Threshold(width As Long, height As Long, lv As Long, hv As Long)
    Dim R As Long
    For x = 0 To width - 1
        For y = 0 To height - 1
            R = Pbytes(2, x, y)
            If R < lv Then R = 0
            If R >= hv Then R = 255
            Pbytes(2, x, y) = R
            Pbytes(1, x, y) = R
            Pbytes(0, x, y) = R
        Next y
    Next x
End Sub


Private Sub Timer1_Timer()
    Timer1.Enabled = False
    SendMessage hwdc, WM_CAP_GET_FRAME, 0, 0
    SendMessage hwdc, WM_CAP_COPY, 0, 0
    Picture1.Picture = Clipboard.GetData
    GetBitmapBits Picture1.Picture, Pinfo.bmWidthBytes * Pinfo.bmHeight, _
        Pbytes(0, 0, 0)
    If Option1.Value = True Then Gray Picture1.ScaleWidth, Picture1.ScaleHeight
    If Option2.Value = True Then
        Gray Picture1.ScaleWidth, Picture1.ScaleHeight
        Lowpass Picture1.ScaleWidth, Picture1.ScaleHeight
    End If
    If Option3.Value = True Then
        Gray Picture1.ScaleWidth, Picture1.ScaleHeight
        Lowpass Picture1.ScaleWidth, Picture1.ScaleHeight
        Contrast Picture1.ScaleWidth, Picture1.ScaleHeight
    End If
    If Option4.Value = True Then
        Gray Picture1.ScaleWidth, Picture1.ScaleHeight
        Lowpass Picture1.ScaleWidth, Picture1.ScaleHeight
        Contrast Picture1.ScaleWidth, Picture1.ScaleHeight
        For i = 1 To Val(Text4.Text)
            Geometricmean Picture1.ScaleWidth, Picture1.ScaleHeight, gms
        Next i
    End If
    If Option5.Value = True Then
        Gray Picture1.ScaleWidth, Picture1.ScaleHeight
        Lowpass Picture1.ScaleWidth, Picture1.ScaleHeight
        Contrast Picture1.ScaleWidth, Picture1.ScaleHeight
        For i = 1 To Val(Text4.Text)
            Geometricmean Picture1.ScaleWidth, Picture1.ScaleHeight, gms
        Next i
```

```
        AdaptiveThreshold Picture1.ScaleWidth, Val(Text2.Text), Val(Text3.Text)
        End If
        SetBitmapBits Picture2.Image, Pinfo.bmWidthBytes * Pinfo.bmHeight, _
            Pbytes(0, 0, 0)
        Picture2.Refresh
        Picture2.Picture = Picture2.Image
        Picture2.Line (Val(Text2.Text) - 2, Val(Text3.Text) - 2)-(Val(Text2.Text) + 2, _
            Val(Text3.Text) + 2), RGB(255, 0, 0), B
         Timer1.Enabled = True
End Sub
```

If the size of the captured image does not fit in the picture box then the image size has to be changed to 160x120 by activating the 'Format' menu. If no webcam is available then a message box will appear with a message "No webcam found!"

5.7 Lesson 6

Objective: To develop a VB6 program to capture and process webcam streaming video for conversion to gray scale image, low-pass filtering, contrast enhancement, geometric-mean filtering, adaptive threshold and a cleaning operation to extract white path and remove unwanted objects from the captured image under varying illumination conditions.

Following steps summarize the program development.

1. All necessary API calls are declared in 'Webcam6.bas' module, same as 'Webcam5.bas', as mentioned in Lesson 5. It is necessary to include this module in 'Form1' of the VB6 program.
2. AVICAP32.DLL is used to capture webcam streaming video through proper API call. The webcam video format should be either RGB24 or YUY2.
3. Under Form1 two 'Picture Box' controls are added, 'Picture1' to capture image as clipboard data from streaming video at a regular interval of 10mS and 'Picture2' to process image from captured image at the same rate with the help of 'Timer1' control.
4. A menu item 'Format' is added in 'Form1' to set the image size to 160x120 pixels.
5. From 'Picture1' image pixel data information is obtained through 'GetObject' API call.
6. Pixel array 'Pbytes(c, x, y)', an 8-bit array, is obtained through 'GetBitmapBits' API call under 'Timer1' control. Each element of 'Pbytes' contains 8-bit RGB color information of each pixel at 'x' and 'y' image co-ordinate. 'c' stands for color; c:2 for red, c:1 for green and c:0 for blue.
7. Pixel array is processed according to option controls 'Option1', 'Option2', 'Option3', 'Option4', 'Option5' or 'Option6'.

8. If 'Option1' is selected then pixel array is processed as gray scale image with the help of procedure 'Gray' and displayed in picture box 'Picture2' through 'SetBitmapBits' API call.

9. If 'Option2' is selected then pixel array is processed first to gray scale image as in step 8 and then low-pass filtered with the help of procedure 'Lowpass' and then displayed in 'Picture2'.

10. If 'Option3' is selected then array is low-pass filtered as in step 9 and then processed for contrast enhancement using histogram stretching technique with the help of procedure 'Contrast' and then displayed in 'Picture2'.

11. If 'Option4' is selected then array is processed for contrast enhancement as in step 10 and then processed for geometric-mean filtering with the help of procedure 'Geometricmean' and then displayed in 'Picture2'. Options are provided for increasing the number of cascaded Geometric-mean filters and the size of mask for each filter.

12. If 'Option5' is selected then an adaptive threshold operation is performed with the help of the procedure 'Adaptive Threshold' and then displayed in 'Picture2'. First the white line width around a reference pixel [at the nominal position (80,110)] is determined with the procedure 'WhiteLineWidth'. If both left and right path width around the reference pixel are found be less than 'MIN_PATH_WIDTH' value then a parameter 'delta' is adjusted to increase the path width by decreasing the threshold value within a range 'delta_max'. Then the procedure 'Threshold' computes new image and the above sequence of operations repeats until a valid white path is obtained.

13. If 'Option6' is selected then an additional cleaning operation is performed to remove unwanted objects with the help of the procedure 'Clean' and then displayed in 'Picture2'.

Option6

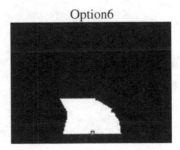

Gray + low-pass + contrast + geometric-mean + threshold + clean

Following figure shows the 'Form1' layout.

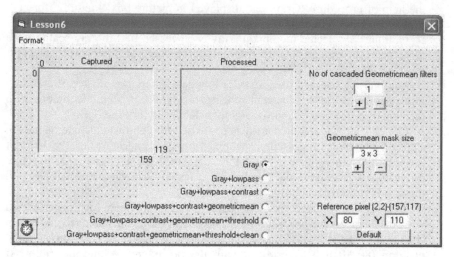

Following text shows the listing of 'Clean' and 'Timer1' procedure code. For rest of the code refer to Lesson 5.

```
Private Sub Clean(width As Long, height As Long, yr As Long)
      Dim R, xr, xref, xwidth As Long
      Dim PB As Long
      Dim bl_flag As Boolean
      bl_flag = False
      xref = 0
      xwidth = 0
      If PixelCountLeft >= MIN_PATH_WIDTH Or PixelCountRight >= _
         MIN_PATH_WIDTH Then
      For x = 0 To width - 1
         R = Pbytes(2, x, yr)
         If R > 240 Then
            If xref = 0 Then xref = x
         End If
         If R > 240 And xref > 0 Then xwidth = xwidth + 1
      Next x
      xr = xref + (xwidth / 2)
      For y = height - 1 To (yr + 1) Step -1
         For x = 0 To width - 1
            Pbytes(2, x, y) = 0
            Pbytes(1, x, y) = 0
            Pbytes(0, x, y) = 0
         Next x
      Next y
      For y = yr To 0 Step -1
         For x = xr To 0 Step -1
```

```
          R = Pbytes(2, x, y)
          If bl_flag = True Then GoTo m1
          If R < 240 Then
             PB = x
             If PB = xr Then bl_flag = True
             GoTo m1
          End If
       Next x
m1:
       For x = PB To 0 Step -1
          Pbytes(2, x, y) = 0
          Pbytes(1, x, y) = 0
          Pbytes(0, x, y) = 0
       Next x

       For x = (xr + 1) To width - 1
          R = Pbytes(2, x, y)
          If bl_flag = True Then GoTo m2
          If R < 240 Then
             PB = x
             If PB = (xr + 1) Then bl_flag = True
             GoTo m2
          End If
       Next x
m2:
       For x = PB To width - 1
          Pbytes(2, x, y) = 0
          Pbytes(1, x, y) = 0
          Pbytes(0, x, y) = 0
       Next x

       xref = 0
       xwidth = 0
       For x = 0 To width - 1
          R = Pbytes(2, x, y)
          If R > 240 Then
             If xref = 0 Then xref = x
          End If
          If R > 240 And xref > 0 Then xwidth = xwidth + 1
       Next x
       If xwidth = 0 Then bl_flag = True
       For x = 0 To width - 1
          If bl_flag = True Then
             Pbytes(2, x, y) = 0
             Pbytes(1, x, y) = 0
             Pbytes(0, x, y) = 0
          End If
```

```
        Next x
      Next y
    End If
End Sub

Private Sub Timer1_Timer()
    Timer1.Enabled = False
    SendMessage hwdc, WM_CAP_GET_FRAME, 0, 0
    SendMessage hwdc, WM_CAP_COPY, 0, 0
    Picture1.Picture = Clipboard.GetData
    GetBitmapBits Picture1.Picture, Pinfo.bmWidthBytes * Pinfo.bmHeight, _
      Pbytes(0, 0, 0)
    If Option1.Value = True Then Gray Picture1.ScaleWidth, Picture1.ScaleHeight
    If Option2.Value = True Then
      Gray Picture1.ScaleWidth, Picture1.ScaleHeight
      Lowpass Picture1.ScaleWidth, Picture1.ScaleHeight
    End If
    If Option3.Value = True Then
      Gray Picture1.ScaleWidth, Picture1.ScaleHeight
      Lowpass Picture1.ScaleWidth, Picture1.ScaleHeight
      Contrast Picture1.ScaleWidth, Picture1.ScaleHeight
    End If
    If Option4.Value = True Then
      Gray Picture1.ScaleWidth, Picture1.ScaleHeight
      Lowpass Picture1.ScaleWidth, Picture1.ScaleHeight
      Contrast Picture1.ScaleWidth, Picture1.ScaleHeight
      For i = 1 To Val(Text4.Text)
        Geometricmean Picture1.ScaleWidth, Picture1.ScaleHeight, gms
      Next i
    End If
    If Option5.Value = True Then
      Gray Picture1.ScaleWidth, Picture1.ScaleHeight
      Lowpass Picture1.ScaleWidth, Picture1.ScaleHeight
      Contrast Picture1.ScaleWidth, Picture1.ScaleHeight
      For i = 1 To Val(Text4.Text)
        Geometricmean Picture1.ScaleWidth, Picture1.ScaleHeight, gms
      Next i
      AdaptiveThreshold Picture1.ScaleWidth, Val(Text2.Text), Val(Text3.Text)
    End If
    If Option6.Value = True Then
      Gray Picture1.ScaleWidth, Picture1.ScaleHeight
      Lowpass Picture1.ScaleWidth, Picture1.ScaleHeight
      Contrast Picture1.ScaleWidth, Picture1.ScaleHeight
      For i = 1 To Val(Text4.Text)
```

```
            Geometricmean Picture1.ScaleWidth, Picture1.ScaleHeight, gms
        Next i
        AdaptiveThreshold Picture1.ScaleWidth, Val(Text2.Text), _
            Val(Text3.Text)
        Clean Picture1.ScaleWidth, Picture1.ScaleHeight, Val(Text3.Text)
    End If
    SetBitmapBits Picture2.Image, Pinfo.bmWidthBytes * Pinfo.bmHeight, _
        Pbytes(0, 0, 0)
    Picture2.Refresh
    Picture2.Picture = Picture2.Image
    Picture2.Line (Val(Text2.Text) - 2, Val(Text3.Text) - 2)-(Val(Text2.Text) _
        + 2, Val(Text3.Text) + 2), RGB(255, 0, 0), B
    Timer1.Enabled = True
End Sub
```

If the size of the captured image does not fit in the picture box then the image size has to be changed to 160x120 by activating the 'Format' menu. If no webcam is available then a message box will appear with a message "No webcam found!"

5.8 Lesson 7

Objective: To develop a VB6 program to capture and process webcam streaming video for conversion to gray scale image, low-pass filtering, contrast enhancement, geometric-mean filtering, adaptive threshold and clean operations along with a selection of white/black path color for vision based navigation.

Following steps summarize the program development.

1. All necessary API calls are declared in 'Webcam7.bas' module, same as 'Webcam6.bas', as mentioned in Lesson 6. It is necessary to include this module in 'Form1' of the VB6 program.
2. AVICAP32.DLL is used to capture webcam streaming video through proper API call. The webcam video format should be either RGB24 or YUY2.
3. Under Form1 two 'Picture Box' controls are added, 'Picture1' to capture image as clipboard data from streaming video at a regular interval of 10mS and 'Picture2' to process image from captured image at the same rate with the help of 'Timer1' control.
4. A menu item 'Format' is added in 'Form1' to set the image size to 160x120 pixels.
5. From 'Picture1' image pixel data information is obtained through 'GetObject' API call.
6. Pixel array 'Pbytes(c, x, y)', an 8-bit array, is obtained through 'GetBitmapBits' API call under 'Timer1' control. Each element of 'Pbytes' contains 8-bit RGB color information of each pixel at 'x' and 'y' image co-ordinate. 'c' stands for color; c:2 for red, c:1 for green and c:0 for blue.

7. 'Shape1' displays the color of the path (white or black) as selected with the 'cmdWhiteBlack' button.
8. Captured image is converted to negative with the help of procedure 'Negative' if black path is selected according to step 7. Then this image is processed according to the option selection ('Option1' to 'Option6') as described in lesson 6.

Following figure shows the 'Form1' layout.

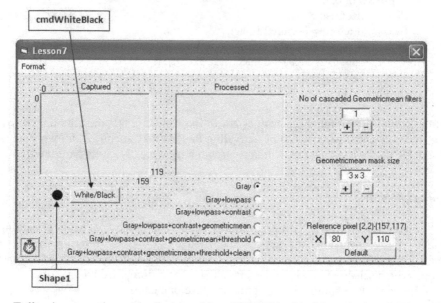

Following text shows the listing of 'cmdWhiteBlack', 'Negative' and 'Timer1' procedure code. For rest of the code refer to Lesson 6.

```
Private Sub cmdWhiteBlack_Click()
    If sflag = False Then
        sflag = True
    Else
        sflag = False
    End If
    If sflag = False Then Shape1.FillColor = vbWhite
    If sflag = True Then Shape1.FillColor = vbBlack
End Sub

Private Sub Negative(width As Long, height As Long)
    Dim R As Long
    For x = 0 To width - 1
        For y = 0 To height - 1
            R = 255 - Pbytes(2, x, y)    'Invert
            Pbytes(2, x, y) = R
            Pbytes(1, x, y) = R
            Pbytes(0, x, y) = R
```

```
        Next y
      Next x
End Sub

Private Sub Timer1_Timer()
  Timer1.Enabled = False
  SendMessage hwdc, WM_CAP_GET_FRAME, 0, 0
  SendMessage hwdc, WM_CAP_COPY, 0, 0
  Picture1.Picture = Clipboard.GetData
  GetBitmapBits Picture1.Picture, Pinfo.bmWidthBytes * Pinfo.bmHeight, _
      Pbytes(0, 0, 0)
  If Option1.Value = True Then
    Gray Picture1.ScaleWidth, Picture1.ScaleHeight
    If sflag = True Then Negative Picture1.ScaleWidth, Picture1.ScaleHeight
  End If
  If Option2.Value = True Then
    Gray Picture1.ScaleWidth, Picture1.ScaleHeight
    If sflag = True Then Negative Picture1.ScaleWidth, Picture1.ScaleHeight
    Lowpass Picture1.ScaleWidth, Picture1.ScaleHeight
  End If
  If Option3.Value = True Then
    Gray Picture1.ScaleWidth, Picture1.ScaleHeight
    If sflag = True Then Negative Picture1.ScaleWidth, Picture1.ScaleHeight
    Lowpass Picture1.ScaleWidth, Picture1.ScaleHeight
    Contrast Picture1.ScaleWidth, Picture1.ScaleHeight
  End If
  If Option4.Value = True Then
    Gray Picture1.ScaleWidth, Picture1.ScaleHeight
    If sflag = True Then Negative Picture1.ScaleWidth, Picture1.ScaleHeight
    Lowpass Picture1.ScaleWidth, Picture1.ScaleHeight
    Contrast Picture1.ScaleWidth, Picture1.ScaleHeight
    For i = 1 To Val(Text4.Text)
      Geometricmean Picture1.ScaleWidth, Picture1.ScaleHeight, gms
    Next i
  End If
  If Option5.Value = True Then
    Gray Picture1.ScaleWidth, Picture1.ScaleHeight
    If sflag = True Then Negative Picture1.ScaleWidth, Picture1.ScaleHeight
    Lowpass Picture1.ScaleWidth, Picture1.ScaleHeight
    Contrast Picture1.ScaleWidth, Picture1.ScaleHeight
    For i = 1 To Val(Text4.Text)
      Geometricmean Picture1.ScaleWidth, Picture1.ScaleHeight, gms
    Next i
    AdaptiveThreshold Picture1.ScaleWidth, Val(Text2.Text), Val(Text3.Text)
  End If
```

```
   If Option6.Value = True Then
      Gray Picture1.ScaleWidth, Picture1.ScaleHeight
      If sflag = True Then Negative Picture1.ScaleWidth, Picture1.ScaleHeight
      Lowpass Picture1.ScaleWidth, Picture1.ScaleHeight
      Contrast Picture1.ScaleWidth, Picture1.ScaleHeight
      For i = 1 To Val(Text4.Text)
         Geometricmean Picture1.ScaleWidth, Picture1.ScaleHeight, gms
      Next i
      AdaptiveThreshold Picture1.ScaleWidth, Val(Text2.Text), _
         Val(Text3.Text)
      Clean Picture1.ScaleWidth, Picture1.ScaleHeight, Val(Text3.Text)
   End If
   SetBitmapBits Picture2.Image, Pinfo.bmWidthBytes * Pinfo.bmHeight, _
      Pbytes(0, 0, 0)
   Picture2.Refresh
   Picture2.Picture = Picture2.Image
   Picture2.Line (Val(Text2.Text) - 2, Val(Text3.Text) - 2) _
      - (Val(Text2.Text) + 2, Val(Text3.Text) + 2), RGB(255, 0, 0), B
   Timer1.Enabled = True
End Sub
```

If the size of the captured image does not fit in the picture box then the image size has to be changed to 160x120 by activating the 'Format' menu. If no webcam is available then a message box will appear with a message "No webcam found!"

5.9 Lesson 8

Objective: To develop a VB6 program to capture and process webcam streaming video for vision based navigation along with a selection of white/black path color. Inference is drawn on whether path is available or not.

Following steps summarize the program development.

1. All necessary API calls are declared in 'Webcam8.bas' module, same as 'Webcam7.bas', as mentioned in Lesson 7. It is necessary to include this module in 'Form1' of the VB6 program.
2. AVICAP32.DLL is used to capture webcam streaming video through proper API call. The webcam video format should be either RGB24 or YUY2.
3. Under Form1 two 'Picture Box' controls are added, 'Picture1' to capture image as clipboard data from streaming video at a regular interval of 10mS and 'Picture2' to process image from the captured image at the same rate with the help of 'Timer1' control.
4. A menu item 'Format' is added in 'Form1' to set the image size to 160x120 pixels.

5. From 'Picture1' image pixel data information is obtained through 'GetObject' API call.
6. Pixel array 'Pbytes(c, x, y)', an 8-bit array, is obtained through 'GetBitmapBits' API call under 'Timer1' control. Each element of 'Pbytes' contains 8-bit RGB color information of each pixel at 'x' and 'y' image co-ordinate. 'c' stands for color; c:2 for red, c:1 for green and c:0 for blue.
7. 'Shape1' displays the color of the path (white or black) as selected with the 'cmdWhiteBlack' button.
8. Captured image is converted to negative with the help of procedure 'Negative' if black path is selected according to step 7. Then this image is processed according to the option 6 of Lesson 7.
9. Then white line width around a fixed reference pixel [at position (80,110)] is determined with the procedure 'WhiteLineWidth'. If both left and right path width around the reference pixel are found be less than 'MIN_PATH_WIDTH' value then 'No path' inference is drawn, otherwise 'Path found' inference is drawn and shown in a text box.

Following figure shows the 'Form1' layout.

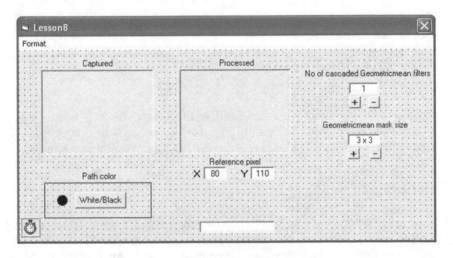

Following text shows the listing of 'Timer1' procedure code. For rest of the code refer to Lesson 7.

```
Private Sub Timer1_Timer()
    Timer1.Enabled = False
    SendMessage hwdc, WM_CAP_GET_FRAME, 0, 0
    SendMessage hwdc, WM_CAP_COPY, 0, 0
    Picture1.Picture = Clipboard.GetData
    GetBitmapBits Picture1.Picture, Pinfo.bmWidthBytes * Pinfo.bmHeight, _
        Pbytes(0, 0, 0)
```

```
  Gray Picture1.ScaleWidth, Picture1.ScaleHeight
  If sflag = True Then Negative Picture1.ScaleWidth, Picture1.ScaleHeight
  Lowpass Picture1.ScaleWidth, Picture1.ScaleHeight
  Contrast Picture1.ScaleWidth, Picture1.ScaleHeight
  For i = 1 To Val(Text4.Text)
      Geometricmean Picture1.ScaleWidth, Picture1.ScaleHeight, gms
  Next i
  AdaptiveThreshold Picture1.ScaleWidth, Val(Text2.Text), Val(Text3.Text)
  Clean Picture1.ScaleWidth, Picture1.ScaleHeight, Val(Text3.Text)
  WhiteLineWidth Picture1.ScaleWidth, Val(Text2.Text), Val(Text3.Text)
  SetBitmapBits Picture2.Image, Pinfo.bmWidthBytes * Pinfo.bmHeight, _
      Pbytes(0, 0, 0)
  Picture2.Refresh
  Picture2.Picture = Picture2.Image
  Picture2.Line (Val(Text2.Text) - 2, Val(Text3.Text) - 2)-(Val(Text2.Text) _
      + 2, Val(Text3.Text) + 2), RGB(255, 0, 0), B
  If PixelCountLeft < MIN_PATH_WIDTH And PixelCountRight < _
      MIN_PATH_WIDTH Then
      Text5.Text = "No path"
  Else
      Text5.Text = "Path found"
  End If
  Timer1.Enabled = True
End Sub
```

If the size of the captured image does not fit in the picture box then the image size has to be changed to 160x120 by activating the 'Format' menu. If no webcam is available then a message box will appear with a message "No webcam found!"

5.10 Lesson 9

Objective: To develop a VB6 program to capture and process webcam streaming video for vision based navigation along with a selection of white/black path color. Inference is drawn on whether path is available or not. Appropriate rules are applied to determine different navigational directions and speed parameters for differential drive.

Following steps summarize the program development.

1. All necessary API calls are declared in 'Webcam9.bas' module, same as 'Webcam8.bas', as mentioned in Lesson 8. It is necessary to include this module in 'Form1' of the VB6 program.
2. AVICAP32.DLL is used to capture webcam streaming video through proper API call. The webcam video format should be either RGB24 or YUY2.

3. Under Form1 two 'Picture Box' controls are added, 'Picture1' to capture image as clipboard data from streaming video at a regular interval of 10mS and 'Picture2' to process image from the captured image at the same rate with the help of 'Timer1' control.

4. A menu item 'Format' is added in 'Form1' to set the image size to 160x120 pixels.

5. From 'Picture1' image pixel data information is obtained through 'GetObject' API call.

6. Pixel array 'Pbytes(c, x, y)', an 8-bit array, is obtained through 'GetBitmapBits' API call under 'Timer1' control. Each element of 'Pbytes' contains 8-bit RGB color information of each pixel at 'x' and 'y' image co-ordinate. 'c' stands for color; c:2 for red, c:1 for green and c:0 for blue.

7. 'Shape1' displays the color of the path (white or black) as selected with the 'cmdWhiteBlack' button.

8. Captured image is processed according to Lesson 8. If path is found then appropriate navigational direction ('forward' or 'turn-left' or 'turn-right') and the corresponding speed parameters for differential drive are determined with three rules. A picture box shows the direction of navigation.

Following figure shows the 'Form1' layout.

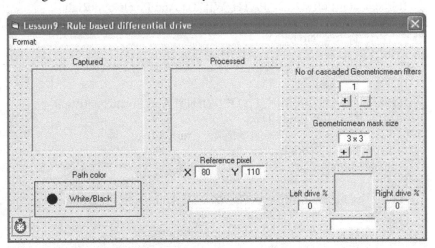

Following text shows the listing of 'Timer1' procedure code. For rest of the code refer to Lesson 8.

```
Private Sub Timer1_Timer()
    Timer1.Enabled = False
    SendMessage hwdc, WM_CAP_GET_FRAME, 0, 0
    SendMessage hwdc, WM_CAP_COPY, 0, 0
    Picture1.Picture = Clipboard.GetData
```

```
GetBitmapBits Picture1.Picture, Pinfo.bmWidthBytes * Pinfo.bmHeight, _
    Pbytes(0, 0, 0)
Gray Picture1.ScaleWidth, Picture1.ScaleHeight
If blkflag = True Then Negative Picture1.ScaleWidth, Picture1.ScaleHeight
Lowpass Picture1.ScaleWidth, Picture1.ScaleHeight
Contrast Picture1.ScaleWidth, Picture1.ScaleHeight
For i = 1 To Val(Text4.Text)
    Geometricmean Picture1.ScaleWidth, Picture1.ScaleHeight, gms
Next i
AdaptiveThreshold Picture1.ScaleWidth, Val(Text2.Text), Val(Text3.Text)
Clean Picture1.ScaleWidth, Picture1.ScaleHeight, Val(Text3.Text)
WhiteLineWidth Picture1.ScaleWidth, Val(Text2.Text), Val(Text3.Text)
SetBitmapBits Picture2.Image, Pinfo.bmWidthBytes * Pinfo.bmHeight, _
    Pbytes(0, 0, 0)
Picture2.Refresh
Picture2.Picture = Picture2.Image
Picture2.Line (Val(Text2.Text) - 2, Val(Text3.Text) - 2)-(Val(Text2.Text) _
    + 2, Val(Text3.Text) + 2), RGB(255, 0, 0), B


If PixelCountLeft < MIN_PATH_WIDTH And PixelCountRight < _
    MIN_PATH_WIDTH Then
    Text5.Text = "No path"
Else
    Text5.Text = "Path found"
End If
If PixelCountLeft >= MIN_PATH_WIDTH And PixelCountRight < _
    MIN_PATH_WIDTH Then
    Text6.Text = 0: Text7.Text = 50        'turn left
    Text8.Text = "Turn left"
    Picture3.Picture = LoadPicture("turn_left.jpg")
End If
If PixelCountLeft < MIN_PATH_WIDTH And PixelCountRight >= _
    MIN_PATH_WIDTH Then
    Text6.Text = 50: Text7.Text = 0        'turn right
    Text8.Text = "Turn right"
    Picture3.Picture = LoadPicture("turn_right.jpg")
End If
If PixelCountLeft >= MIN_PATH_WIDTH And PixelCountRight >= _
    MIN_PATH_WIDTH Then
    Text6.Text = 100: Text7.Text = 100     'forward
    Text8.Text = "Forward"
    Picture3.Picture = LoadPicture("forward.jpg")
End If
```

```
    If PixelCountLeft < MIN_PATH_WIDTH And PixelCountRight < _
    MIN_PATH_WIDTH Then
        Text6.Text = 0: Text7.Text = 0      'no path - idle
        Text8.Text = ""
        Picture3.Picture = LoadPicture("blank.jpg")
    End If
    Timer1.Enabled = True
End Sub
```

Following image files are used to indicate direction of navigation.

Forward.jpg   turn_left.jpg   turn_right.jpg

If the size of the captured image does not fit in the picture box then the image size has to be changed to 160x120 by activating the 'Format' menu. If no webcam is available then a message box will appear with a message "No webcam found!"

5.11 Lesson 10

Objective: To develop a VB6 program to capture and process webcam streaming video for vision based navigation along with a selection of white/black path color. Inference is drawn on whether path is available or not. Appropriate rules are applied to determine different navigational directions and speed parameters for differential drive. Sound output is added to draw attention.

Following steps summarize the program development.

1. All necessary API calls are declared in 'Webcam10.bas' module. It is necessary to include this module in 'Form1' of the VB6 program.
2. AVICAP32.DLL is used to capture webcam streaming video through proper API call. The webcam video format should be either RGB24 or YUY2.
3. Under Form1 two 'Picture Box' controls are added, 'Picture1' to capture image as clipboard data from streaming video at a regular interval of 10mS and 'Picture2' to process image from the captured image at the same rate with the help of 'Timer1' control.
4. A menu item 'Format' is added in 'Form1' to set the image size to 160x120 pixels.
5. From 'Picture1' image pixel data information is obtained through 'GetObject' API call.
6. Pixel array 'Pbytes(c, x, y)', an 8-bit array, is obtained through 'GetBitmapBits' API call under 'Timer1' control. Each element of 'Pbytes' contains 8-bit RGB color information of each pixel at 'x' and 'y' image co-ordinate. 'c' stands for color; c:2 for red, c:1 for green and c:0 for blue.

7. 'Shape1' displays the color of the path (white or black) as selected with the 'cmdWhiteBlack' button.
8. Captured image is processed according to Lesson 9. If path is found then appropriate navigational direction ('forward' or 'turn-left' or 'turn-right') and the corresponding speed parameters for differential drive are determined with three rules. A picture box shows the direction of navigation.
9. Sound output is activated through 'sndPlaySound' API call with appropriate 'wave' file.

Following text shows the listing of 'Webcam10.bas' module.

```
Global Const WS_CHILD As Long = &H40000000
Global Const WS_VISIBLE As Long = &H10000000
Global Const WM_USER = 1024
Global Const WM_CAP_DRIVER_CONNECT = WM_USER + 10
Global Const WM_CAP_SET_PREVIEW = WM_USER + 50
Global Const WM_CAP_SET_PREVIEWRATE = WM_USER + 52
Global Const WM_CAP_DRIVER_DISCONNECT As Long = WM_USER + 11
Global Const WM_CAP_DLG_VIDEOFORMAT As Long = WM_USER + 41
Global Const WM_CAP_DLG_VIDEOCOMPRESSION As Long = _
    WM_USER + 46
Global Const WM_CAP_DLG_VIDEODISPLAY As Long = WM_USER + 43
Global Const WM_CAP_DLG_VIDEOSOURCE As Long = WM_USER + 42
Global Const WM_CAP_GET_FRAME As Long = 1084
Global Const WM_CAP_COPY As Long = 1054
Global Const WM_CAP_SET_SCALE As Integer = WM_USER + 53
Global Const SWP_NOMOVE As Integer = 2
Global Const SWP_NOZORDER As Integer = 4
Global Const HWND_BOTTOM As Integer = 1
Global Const SND_ASYNC = 1
Global Const SND_LOOP = 8
Global Const SND_NODEFAULT = 2
Global Const SND_SYNC = 0
Global Const SND_NOSTOP = 16
Global Const SND_MEMORY = 4


Declare Function SendMessage Lib "user32" Alias "SendMessageA" (ByVal hwnd _
    As Long, ByVal wMsg As Long, ByVal wParam As Long, ByVal lParam As Long) _
    As Long Declare Function capCreateCaptureWindow Lib "avicap32.dll" Alias _
    "capCreateCaptureWindowA" (ByVal nWindowName As String, ByVal nStyle _
    As Long, ByVal nx As Integer, ByVal ny As Integer, ByVal nWidth As Integer, _
    ByVal nHeight As Integer, ByVal nHwnd As Long, ByVal nId As Integer) As Long
Declare Function SetWindowPos Lib "user32" (ByVal hwnd As Long, _
    ByVal hWndInsertAfter As Long, ByVal x As Long, ByVal y As Long, _
    ByVal cx As Long, ByVal cy As Long, ByVal wFlags As Long) As Long
```

Declare Function GetObject Lib "gdi32" Alias "GetObjectA" (ByVal hObject As Long, _
    ByVal nCount As Long, lpObject As Any) As Long
Declare Function GetBitmapBits Lib "gdi32" (ByVal hBitmap As Long, ByVal dwCount _
    As Long, lpBits As Any) As Long
Declare Function SetBitmapBits Lib "gdi32" (ByVal hBitmap As Long, ByVal dwCount _
    As Long, lpBits As Any) As Long
Declare Function sndPlaySound Lib "winmm.dll" Alias "sndPlaySoundA" _
    (ByVal lpszSoundName As String, ByVal uFlags As Long) As Long

Following figure shows the 'Form1' layout.

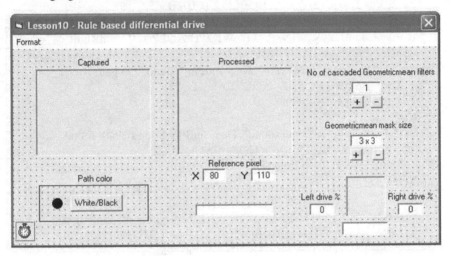

Following text shows the listing of 'Timer1' procedure code. For rest of the
code refer to Lesson 9.

```
Private Sub Timer1_Timer()
    Timer1.Enabled = False
    SendMessage hwdc, WM_CAP_GET_FRAME, 0, 0
    SendMessage hwdc, WM_CAP_COPY, 0, 0
    Picture1.Picture = Clipboard.GetData
    GetBitmapBits Picture1.Picture, Pinfo.bmWidthBytes * Pinfo.bmHeight, _
        Pbytes(0, 0, 0)

    Gray Picture1.ScaleWidth, Picture1.ScaleHeight
    If blkflag = True Then Negative Picture1.ScaleWidth, Picture1.ScaleHeight
    Lowpass Picture1.ScaleWidth, Picture1.ScaleHeight
    Contrast Picture1.ScaleWidth, Picture1.ScaleHeight
    For i = 1 To Val(Text4.Text)
        Geometricmean Picture1.ScaleWidth, Picture1.ScaleHeight, gms
    Next i
```

```
AdaptiveThreshold Picture1.ScaleWidth, Val(Text2.Text), Val(Text3.Text)
Clean Picture1.ScaleWidth, Picture1.ScaleHeight, Val(Text3.Text)
WhiteLineWidth Picture1.ScaleWidth, Val(Text2.Text), Val(Text3.Text)

SetBitmapBits Picture2.Image, Pinfo.bmWidthBytes * Pinfo.bmHeight, _
    Pbytes(0, 0, 0)
Picture2.Refresh
Picture2.Picture = Picture2.Image
Picture2.Line (Val(Text2.Text) - 2, Val(Text3.Text) - 2)-(Val(Text2.Text) _
    + 2, Val(Text3.Text) + 2), RGB(255, 0, 0), B

If PixelCountLeft < MIN_PATH_WIDTH And PixelCountRight < _
    MIN_PATH_WIDTH Then
    If Text5.Text <> "No path" Then sndPlaySound "No path.wav", _
    SND_ASYNC Or SND_NODEFAULT
    Text5.Text = "No path"
Else
    If Text5.Text <> "Path found" Then sndPlaySound "Path found.wav", _
        SND_ASYNC Or SND_NODEFAULT
    Text5.Text = "Path found"
End If


If PixelCountLeft >= MIN_PATH_WIDTH And PixelCountRight < _
    MIN_PATH_WIDTH Then
    Text6.Text = 0: Text7.Text = 50      'turn left
    Text8.Text = "Turn left"
    Picture3.Picture = LoadPicture("turn_left.jpg")
End If
If PixelCountLeft < MIN_PATH_WIDTH And PixelCountRight >= _
    MIN_PATH_WIDTH Then
    Text6.Text = 50: Text7.Text = 0      'turn right
    Text8.Text = "Turn right"
    Picture3.Picture = LoadPicture("turn_right.jpg")
End If
If PixelCountLeft >= MIN_PATH_WIDTH And PixelCountRight >= _
    MIN_PATH_WIDTH Then
    Text6.Text = 100: Text7.Text = 100     'forward
    Text8.Text = "Forward"
    Picture3.Picture = LoadPicture("forward.jpg")
End If
```

```
    If PixelCountLeft < MIN_PATH_WIDTH And PixelCountRight < _
    MIN_PATH_WIDTH Then
    Text6.Text = 0: Text7.Text = 0     'no path - idle
    Text8.Text = ""
    Picture3.Picture = LoadPicture("blank.jpg")
    End If
    Timer1.Enabled = True
  End Sub
```

Two pre-recorded wave files 'Nopath.wav' and 'Pathfound.wav' are used to play when needed through PC sound card interface. The PC sound recorder program may be used to create these wave files.

If the size of the captured image does not fit in the picture box then the image size has to be changed to 160x120 by activating the 'Format' menu. If no webcam is available then a message box will appear with a message "No webcam found!"

5.12 Summary

Ten lessons are presented in a step-by-step manner to develop programming skill for implementing vision-based navigation applications under 32-bit Windows environment.

Lesson 1: This demonstrates how to capture image frames from streaming video from a low-cost webcam and examine pixel values with the help of mouse pointer.

Lesson 2: This demonstrates how to process captured image frames from streaming video with two processing options covering color to gray-scale conversion and low-pass filtering.

Lesson 3: The method of contrast enhancement by histogram stretching technique is added to improve contrast under poor lighting conditions.

Lesson 4: The geometric-mean filter is added to smooth and suppress image detail.

Lesson 5: An adaptive threshold operation in introduced to extract white path under varying illumination conditions.

Lesson 6: A cleaning operation is provided to remove unwanted objects detected.

Lesson 7: Here an option is added for selection of path color white or black.

Lesson 8: Modified for white or black path searching for navigation with reference to a fixed pixel.

Lesson 9: Introduces a rule-based approach to determine left and right wheel speed settings of a differential drive system for navigation.

Lesson 10: Here sound output is added to draw attention during navigation.

References

[1] Balena, F.: Programming Microsoft Visual Basic 6. Microsoft Press (1999)
[2] Mandelbrot Set International Ltd., Advanced Microsoft Visual Basic 6. Microsoft Press (1998)
[3] Appleman, D.: Dan Appleman's Win32 API Puzzle Book and Tutorial for Visual Basic Programmers. Apress (1999)
[4] Gonzalez, Woods: Digital Image Processing. Prentice Hall (2002)

Chapter 6
Vision Based Mobile Robot Path/Line Tracking

Abstract. In this chapter we discuss how a vision based navigation scheme can be developed for indoor path/line tracking, so that the robot is equipped to follow a narrow line or to travel along a wide path. The scheme utilizes fuzzy logic to achieve the desired objective. The scheme is so developed that, in case of absence of obstacles in front, it will guide the robot to navigate using fuzzy vision-based navigation. The scheme also employs a fuzzy IR-based obstacle avoidance strategy which becomes active on detection of any obstacle.

6.1 Introduction

In this chapter we shall describe a vision-based navigation algorithm implemented in conjunction with the robot indigenously developed in our laboratory, which utilizes fuzzy logic for path/line tracking, in presence or in absence of obstacle [10]. Fuzzy logic has been widely accepted as a possible means in mobile robot navigation for quite some time now. In [1], an earliest fuzzy controller was developed for obstacle avoidance. In that work the controller used a vision based algorithm to obtain information about occupied and free areas in front of the robot from a video camera and the rules were derived with the help of a simulator. Another similar work for corridor navigation, by a fuzzy controller, using video images, which was implemented in vehicle ATHENE, was reported in [2]. Fuzzy logic approaches have been widely utilized in navigation systems for mobile robots over a decade. A method of path planning and execution, using fuzzy logic, for mobile robot control, was proposed in [3]. Almost during similar time, a successful application of fuzzy logic for vision based mobile robot navigation, considering the aspects of collision avoidance and obstacle avoidance, was reported in [4]. In [5], a new approach based on forecast learning fuzzy control, where the environmental information is acquired by a CCD camera, was proposed. In this work the image acquired is classified into several characteristic patterns and the robot is programmed with sets of control rules for each pattern, set *a priori*. The robot combines these sets into a single set by matching the patterns. Several works have also been reported with stereo vision system, coupled with the support of conventional sensors, for robot navigation, using fuzzy controllers [6], [7]. A detailed and comprehensive study of several fuzzy based mobile robot

A. Chatterjee et al.: Vision Based Autonomous Robot Navigation, SCI 455, pp. 143–166.
springerlink.com   © Springer-Verlag Berlin Heidelberg 2013

navigation techniques was presented in [8]. In recent times, a new fuzzy based approach called rule-based fuzzy traversability index approach is used for outdoor navigation of mobile robots, where imagery data is used to classify different characteristics like terrain roughness, terrain slope, terrain discontinuity, terrain hardness etc. [9]. Once these characteristics from the viewable scene are extracted, then the fuzzy rules for traversablility index are developed for smooth navigation of the mobile robot.

Utilizing the indigenous robot developed in our laboratory as described in the previous chapter, a new fuzzy based mobile robot navigation scheme is developed which attempts to track the middle of a narrow line or a broad path, both in presence or absence of obstacle. This system utilizes a vision-based fuzzy module for navigation when there is no obstacle in front of the robot. As soon as the robot senses an obstacle in front, it deactivates the vision-based fuzzy module and activates an IR-based fuzzy obstacle avoidance module, so that the robot attempts to safely avoid the obstacle and re-localize itself on the middle of the path/line. If this objective is satisfied, then the IR-based fuzzy module is deactivated and vision-based fuzzy module is re-activated and the robot continues with its line tracking activity. The robot system utilizes the capability of intranet-connectivity, suitable for client/server operation, as described in the previous chapter, so that the robot functionalities can be suitably chosen and the robot can be suitably commanded from a remote end client PC.

6.2 A Preview of the Proposed Scheme

Figure 6.1 shows the complete scheme developed in this work. Let the pose of the differential drive robot system, at the present given instant, be (x_R, y_R, ϕ_R). Depending on the environment ahead of it, a new navigation command is issued for the robot that comprises the linear velocity command (v) and the steering angle command (θ). The steering angle command can be any value between (0^0-180^0), counted in a counter-clockwise sense, with reference to the present pose of the robot. This is shown in Fig. 6.2 where the World Coordinate System is denoted by *XWY* and the mobile robot coordinate system is denoted by *xoy*, *o* being the center of the robot. The new direction of the robot navigation is along *op* in Fig. 6.2. At any given position, the robot scans the front of it, using the IR sensor at positions 4, 5, and 6, to determine whether the front region is free from obstacle or it contains an obstacle. If the presence of an obstacle is detected, it will first produce a voice message that there is an obstacle in front, hoping that somebody has wrongly left an obstacle in its path and will remove it, hearing the robot speak. If this does not happen, the robot will perform the obstacle avoidance using the IR sensor readings in eleven scan positions. The obstacle avoidance routine will be so performed that the robot will attempt to take a short detour in its original path and, after avoiding the obstacle, it will attempt to come back to its original, ideal path.

Once the robot re-detects that there is no obstacle in front, the system will return the control to its vision-based navigation scheme. The system developed employs one fuzzy based navigation algorithm each, for both vision-based navigation and IR-based obstacle avoidance. The basic philosophy of the navigation scheme is that the robot should track the center of a path towards its goal, whether in presence or in absence of any obstacle in its path. For a wide path, the robot always attempts to navigate through the middle of the path. Similarly, for a narrow path or line, whose width is smaller than the width of the robot, the navigation algorithm attempts to track the center of the line.

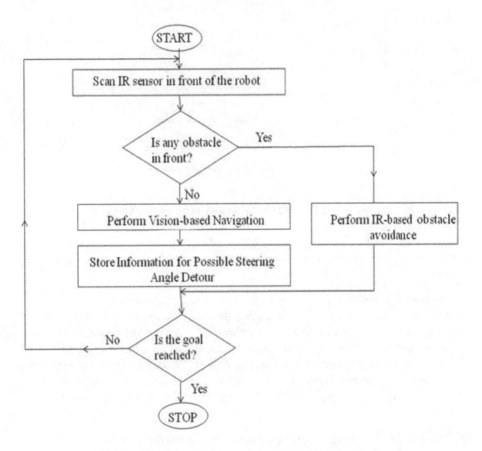

Fig. 6.1. The navigation strategy for the mobile robot

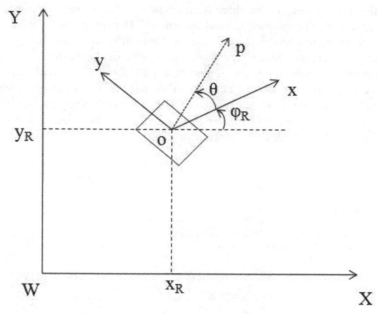

Fig. 6.2. The robot co-ordinate system

6.3   A Fuzzy System for Vision Based Robot Navigation

If the robot find that its front region is free of obstacle, the robot will undertake vision based navigation. The scheme employs the following image processing steps:

a. Capture a frame from the video stream recorded by the webcam
When the IR scanning system infers that the front region of the robot is a free region, a frame is captured from the continuously running video stream available from the webcam of the laptop, mounted on the robot. This frame gives the visual information of the environment ahead of the robot. This frame is further processed to extract meaningful information from it, by first converting the colour image to its corresponding gray image and performing image processing steps on this gray image.

b. Process the gray image of the environment to extract the path/line
The next step is carried out to segment the image so that the path/line is extracted from its surroundings. For this the image is first de-speckled to perform low pass filtering, to eliminate noises. Then the image is auto corrected for its brightness, so that, if the image looks unsatisfactorily dark, because of dim illumination, the overall brightness of the processed image can be enhanced by changing the dynamic range of the intensity values. This brightness corrected image is then processed so that the isolated bright spots get connected and thickened, in an operation very similar to dilation by a structuring element. This linking and

thickening operation can be performed by employing a geometric mean filtering technique.

This thickened image is finally segmented by performing thresholding. The intensity threshold is chosen as a very high value in a bid to extract only the path/line from its surrounding. Figure 6.3 and Fig. 6.4 show a sample environment with the output of each image processing step described above, without incorporating the geometric mean filtering step and with incorporation of the geometric mean filtering step. Figure 6.5 and Fig. 6.6 show the similar situations in an environment where there is an interfering object apart from the actual path/line, in the captured image. Figure 6.6 shows how the geometric mean filtering process helps to remove that interfering object through the segmentation process and can clearly extract the path, which was not possible in Fig. 6.5.

c. Employ the fuzzy system for vision-based navigation
The fuzzy-based system is developed based on the thresholded image obtained in the previous step. In this case the image is of size 160 x 120 where the top left corner pixel is assigned the coordinate (0, 0) and the bottom right corner pixel is assigned the coordinate (159, 119). Then a seed point S is chosen on the mid-vertical line on the image, more towards the bottom of the image i.e. corresponding to a real-world point closer to the robot, in its present position. In image pixel coordinates this seed point is chosen as (80, 110). At this position a horizontal line is drawn on the image. From the seed point S, one can travel along this scan line, once towards left and once towards right, to compute the number of pixels (both to the left and to the right of S) with bright intensity, in a bid to

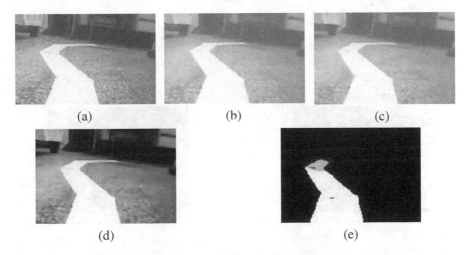

(a)   (b)   (c)

(d)   (e)

Fig. 6.3. The results of the image processing steps for a sample environment: (a) the original image captured, (b) the corresponding gray image, (c) de-speckled image, (d) auto-brightness corrected image, and (e) final processed image after thresholding

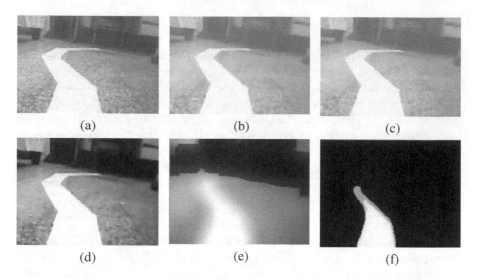

Fig. 6.4. The results of the image processing steps for the sample environment in Fig. 6.3: (a) the original image captured, (b) the corresponding gray image, (c) de-speckled image, (d) auto-brightness corrected image, (e) isolated point linked and thickened image employing geometric mean, and (e) final processed image after thresholding

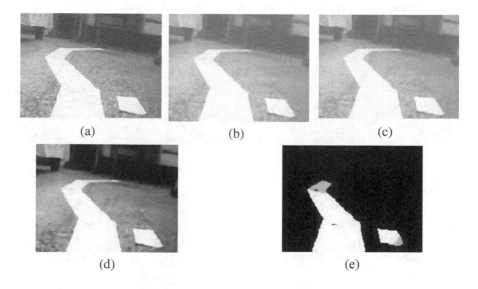

Fig. 6.5. The results of the image processing steps for a sample environment with an interfering object: (a) the original image captured, (b) the corresponding gray image, (c) de-speckled image, (d) auto-brightness corrected image, and (e) final processed image after thresholding

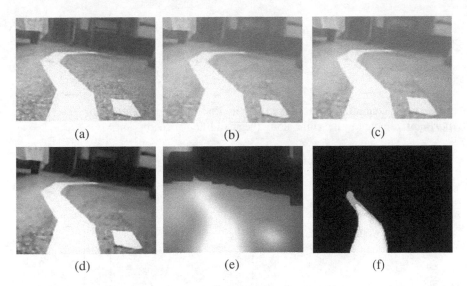

(a)   (b)   (c)

(d)   (e)   (f)

Fig. 6.6. The results of the image processing steps for a sample environment with an interfering object: (a) the original image captured, (b) the corresponding gray image, (c) de-speckled image, (d) auto-brightness corrected image, (e) isolated point linked and thickened image employing geometric mean, and (e) final processed image after thresholding

determine the width of the path towards the left and towards the right of the robot. If these two pixel counts are same, one can infer that the robot is positioned approximately in the middle of the road or the line. On the other hand, if the left pixel count is higher than the right pixel count, it indicates that the robot position is more skewed towards the right of the path and the fuzzy inference system should try to direct it towards the middle of the road. If the right pixel count is higher than the left pixel count, it indicates that the robot is more towards the left side of the road and the fuzzy guidance should again provide a different command to bring the robot back to the middle of the road. This fuzzy-logic based vision-aided navigation system is developed as a two-input-two-output system, where the two input variables are *PixelCountLeft* and *PixelCountRight* and the two output variables are the linear velocity command (v) and the steering angle command (θ). The fuzzy system is developed as a zero-order Takagi-Sugano (TS) system. To make the fuzzy system a robust one, the pixel counts to the left and to the right of the seed point are taken for three consecutive horizontal lines drawn at three seed points ($S_1 \equiv (80, 109)$, $S_2 \equiv (80, 110)$, and $S_3 \equiv (80, 111)$) and then an average count is used as:

$$PixelCount\;\;Left\;\;=\frac{1}{3}\sum_{i=1}^{3}pcl_{i} \qquad (6.1a)$$

$$PixelCount\;\;Right\;\;=\frac{1}{3}\sum_{i=1}^{3}pcr_{i} \qquad (6.1b)$$

where

pcl_i = pixel count to the left along the scan line for seed point S_i and
pcr_i = pixel count to the right along the scan line for seed point S_i.

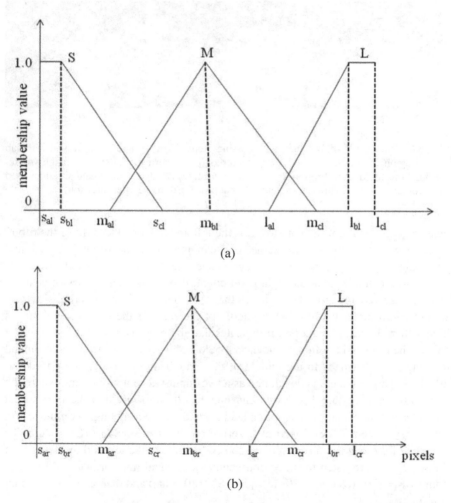

Fig. 6.7. Membership functions for (a) *PixelCountLeft* and (b) *PixelCountRight*

Figure 6.7 shows the input membership functions (MFs) for fuzzification, where each input variable is fuzzified using three MFs: small (S), medium (M), and large (L). The fuzzy sets or MFs for the input variables are described as:

$$\mu_S\left(PixelCountLeft\right) = \begin{cases} \left(\dfrac{s_{cl} - PixelCountLeft}{s_{cl} - s_{bl}}\right), & s_{bl} < PixelCountLeft < s_{cl} \\ 1, & 0 \le PixelCountLeft \le s_{bl} \\ 0, & \text{otherwise} \end{cases} \quad (6.2a)$$

$$\mu_M\left(PixelCountLeft\right) = \begin{cases} \left(\dfrac{PixelCountLeft - m_{al}}{m_{bl} - m_{al}}\right), & m_{al} < PixelCountLeft < m_{bl} \\ \left(\dfrac{m_{cl} - PixelCountLeft}{m_{cl} - m_{bl}}\right), & m_{bl} \le PixelCountLeft < m_{cl} \\ 0, & \text{otherwise} \end{cases} \quad (6.2b)$$

$$\mu_L\left(PixelCountLeft\right) = \begin{cases} \left(\dfrac{PixelCountLeft - l_{al}}{l_{bl} - l_{al}}\right), & l_{al} < PixelCountLeft < l_{bl} \\ 1, & PixelCountLeft \ge l_{bl} \\ 0, & \text{otherwise} \end{cases} \quad (6.2c)$$

$$\mu_S\left(PixelCountRight\right) = \begin{cases} \left(\dfrac{s_{cr} - PixelCountRight}{s_{cr} - s_{br}}\right), & s_{br} < PixelCountRight < s_{cr} \\ 1, & 0 \le PixelCountRight \le s_{br} \\ 0, & \text{otherwise} \end{cases} \quad (6.3a)$$

$$\mu_M\left(PixelCountRight\right) = \begin{cases} \left(\dfrac{PixelCountRight - m_{ar}}{m_{br} - m_{ar}}\right), & m_{ar} < PixelCountRight < m_{br} \\ \left(\dfrac{m_{cr} - PixelCountRight}{m_{cr} - m_{br}}\right), & m_{br} \le PixelCountRight < m_{cr} \\ 0, & \text{otherwise} \end{cases} \quad (6.3b)$$

$$\mu_L\left(PixelCountRight\right) = \begin{cases} \left(\dfrac{PixelCountRight - l_{ar}}{l_{br} - l_{ar}}\right), & l_{ar} < PixelCountRight < l_{br} \\ 1, & PixelCountRight \ge l_{br} \\ 0, & \text{otherwise} \end{cases} \quad (6.3c)$$

The outputs are represented by singletons, for each output variable. The fuzzy rule base consists of a collection of fuzzy **IF-THEN** rules. A model rule i can be given as:

$$R_{vis}^{(i)} : \text{IF} \quad x_1 \text{ is } M_{i1} \text{ AND} \quad x_2 \text{ is } M_{i2} \tag{6.4}$$

$$\text{THEN} \quad y_1 \text{ is } V_{vis\_i1} \text{ AND} \quad y_2 \text{ is } \theta_{vis\_i2}, \quad i = 1, 2, \cdots, N$$

where

$$\mathbf{x} = [x_1, x_2]^T = [PixelCountLeft, PixelCountRight]^T,$$

$$\mathbf{y} = [y_1, y_2]^T = [v_{vis}, \theta_{vis}]^T,$$

$$M_{i1} \in \{S, M, L\}, M_{i2} \in \{S, M, L\}, V_{vis\_i1} \in \mathbf{v}_{vis} \text{ and } \theta_{vis\_i2} \in \mathbf{\theta}_{vis}.$$

Here,

$\mathbf{V}_{vis}$ = vector of singletons for the output linear velocity = $[v_{vis1}, v_{vis2}, \cdots, v_{visN}]^T$ and

$\mathbf{\theta}_{vis}$ = vector of singletons for the output steering angle = $[\theta_{vis1}, \theta_{vis2}, \cdots, \theta_{visN}]^T$.

Table 6.1. Fuzzy rule base for the vision system

Rule No.	Antecedent Parts (IF clauses)		Consequence Parts (THEN Parts)	
	PixelCountLeft	PixelCountRight	v_{vis} (in p.u.)	θ_{vis} (in degree)
1	Small	Small	0.9	90
2	Small	Medium	0.5	67
3	Small	Large	0.1	45
4	Medium	Small	0.5	112
5	Medium	Medium	0.9	90
6	Medium	Large	0.5	67
7	Large	Small	0.1	135
8	Large	Medium	0.5	112
9	Large	Large	0.9	90

N is the total number of rules in the fuzzy rule base. This fuzzy rule base constructed is given in Table 6.1. The fuzzy output for linear velocity is generated in p.u., which is multiplied by a suitable gain (K_{vel_vis}). The defuzzification is carried out by employing weighted average method. Then the output crisp linear velocity command (v_{vis}) and the output steering angle command (θ_{vis}) are computed as:

$$v_{vis} = (K_{vel\_vis}) * \left(\frac{\sum_{i=1}^{N} v_{visi} * \alpha_i(\mathbf{x})}{\sum_{i=1}^{N} \alpha_i(\mathbf{x})} \right) \tag{6.5}$$

$$\theta_{vis} = \frac{\sum_{i=1}^{N} \theta_{visi} * \alpha_i(\mathbf{x})}{\sum_{i=1}^{N} \alpha_i(\mathbf{x})} \tag{6.6}$$

where $\alpha_i(\mathbf{x})$ = firing degree of rule $i = \prod_{j=1}^{2} \mu_i(x_j)$.

(d) Store the possible steering angle detour, if the robot leave the line
The robot navigation system is equipped with an additional module to take care of an excigency situation. Let us consider that, under some circumstances, the robot leaves the path and, from the processed image output, the *PixelCountLeft* and *PixelCountRight* variables are both computed as zero. In this situation the robot is given small steering angle detour commands (with linear velocity chosen as zero), in an iterative fashion, until at least one of the variables *PixelCountLeft* and *PixelCountRight* gives a non-zero count. Then one can infer that the robot has been oriented back to the original path and hence the subsequent activation of the vision-based navigation algorithm will attempt to bring the robot back on the middle of the path/line. Now, whether the robot detour should be activated in clockwise or counter-clockwise direction, can be determined on the basis of whether the robot was moving more towards its left or more towards its right in its previous iterations. This can help in reducing the time to be spent in the detour phase and also to restore the original direction of navigation.

Fig. 6.8 shows the algorithm for storing information for possible steering angle detour. At each sampling instant (k), calculate the number of pixels to the left (pcl_k) and to the right (pcr_k) of a seed point by making a horizontal scan, to determine the number of bright pixels. To determine a proper trend of robot orientation, this process is repeated for a number of rows (N_rows) to determine the cumulative values at the sampling instant k as cum_pcl_k and cum_pcr_k. This process is repeated for each processed image frame in vision-based navigation to determine final stored values of these two corresponding quantities at instant k. However, while storing these values, the highest priority is given to the present instant and as we go back to the past instants, the priority gradually reduces. This can be formulated as:

$$store\_pcl_k = k_1 * cum\_pcl_k + k_2 * store\_pcl_{k-1} \tag{6.7}$$

$$store\_pcr_k = k_1 * cum\_pcr_k + k_2 * store\_pcr_{k-1} \tag{6.8}$$

In this system, the forgetting factor is so chosen that $k_1 = 0.25$ and $k_2 = 0.75$. When the vision-based navigation algorithm is working satisfactorily, the storage continues. However these stored values only become functional when, due to some reason, the robot leaves the path/line and both the *PixelCountLeft* and *PixelCountRight* variables become zero. Then, depending on the polarity of $(store\_pcl_k - store\_pcr_k)$, the steering detour direction to be effected is chosen.

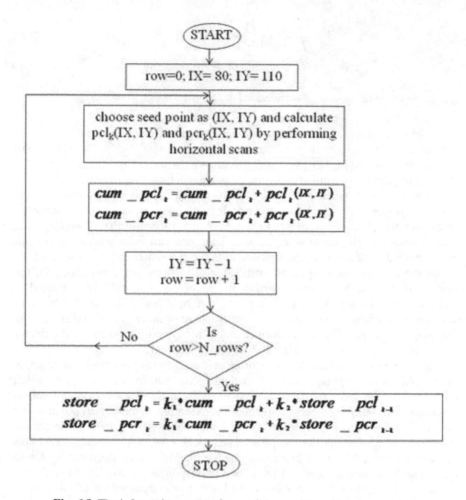

Fig. 6.8. The information storage for possible steering angle detour module

If this quantity is positive, the detour direction is chosen counter-clockwise, otherwise clockwise. Once the detour direction is fixed, an iterative procedure is implemented, where, with zero linear velocity, the robot turns by a fixed angle of 10^0, an image frame is captured and the image processing steps discussed in the previous section are implemented, to determine the new values of the variables *PixelCountLeft* and *PixelCountRight*. If at least one of these values is non-zero, the vision-based navigation algorithm is reactivated. Otherwise, the next iteration of turning the robot by 10^0 and implementing subsequent steps is carried out and this process continues, until the vision-based navigation algorithm gets reactivated.

6.4   The IR-Sensor Based Obstacle Avoidance by Employing a Fuzzy Algorithm

The IR-sensor based obstacle avoidance module will be activated if the system detects an obstacle in front and then the vision system will be deactivated. A fuzzy based IR-obstacle-avoidance scheme will attempt that the robot should go around the obstacle and then continue along its original path. Once the robot avoids the obstacle, then the vision-based algorithm will be reactivated. This will automatically bring the robot back to the middle of the path. For the development of the IR-based fuzzy system, the lone IR-sensor is scanned in eleven angular positions $l = 1, 2, \cdots, 11$ to produce eleven IR sensor readings, $IR\_Sensor\_val(l)$, given in terms of voltage (V). These eleven readings are grouped into three sensor groups $IR\_Group\_val(p), p=1,2,3$. This is done with an aim to reduce the input dimension for the fuzzy system developed. In each IR sensor group, the maximum sensor scan reading is chosen as the representative reading for the group. This is because a higher reading indicates presence of a nearer obstacle. Hence these analog group readings are given as:

$$IR\_Group\_val(1) = \max\left(IR\_Sensor\_val(l)\mid l = 1, 2, 3, 4\right) \quad (6.9)$$

$$IR\_Group\_val(2) = \max\left(IR\_Sensor\_val(l)\mid l = 5, 6, 7\right) \quad (6.10)$$

$$IR\_Group\_val(3) = \max\left(IR\_Sensor\_val(l)\mid l = 8, 9, 10, 11\right) \quad (6.11)$$

Then a three-input-two-output fuzzy obstacle avoidance system is developed with $IR\_Group\_val(p), p = 1,2,3$, as the three inputs and $(v,\ \theta)$ as the two outputs. Here also a zero- order Takagi-Sugeno (TS) fuzzy system is developed. Figure 6.9 shows the MFs chosen for each input variable. Each input is fuzzified using three MFs: far (FR), intermediate (IM), and near (NR). The corresponding MFs can be given as:

$$\mu_{FR}(IR\_Group\_val(p)) = \begin{cases} \left(\dfrac{F_{p3} - IR\_Group\_val(p)}{F_{p3} - F_{p2}}\right), & F_{p2} < IR\_Group\_val(p) < F_{p3} \\ 1, & 0 \leq IR\_Group\_val(p) \leq F_{p2} \\ 0, & otherwise \end{cases}$$

$$(6.12a)$$

$$\mu_{IM}(IR\_Group\_val(p)) = \begin{cases} \left(\dfrac{IR\_Group\_val(p) - I_{p1}}{I_{p2} - I_{p1}}\right), & I_{p1} < IR\_Group\_val(p) < I_{p2} \\ \left(\dfrac{I_{p3} - IR\_Group\_val(p)}{I_{p3} - I_{p2}}\right), & I_{p2} \leq IR\_Group\_val(p) < I_{p3} \\ 0, & otherwise \end{cases} \quad (6.12b)$$

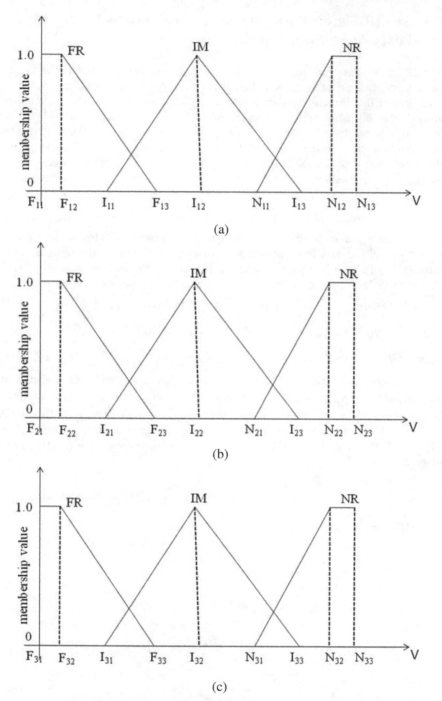

Fig. 6.9. Membership functions for (a) *IR_Group_val*(1), (b) *IR_Group_val*(2), and (c) *IR_Group_val*(3)

$$\mu_{NR}(IR\_Group\_val(p)) = \begin{cases} \left(\dfrac{IR\_Group\_val(p) - N_{p1}}{N_{p2} - N_{p1}}\right), & N_{p1} < IR\_Group\_val(p) < N_{p2} \\ 1, & IR\_Group\_val(p) \geq N_{p2} \\ 0, & otherwise \end{cases}$$

$$(6.12c)$$

Each fuzzy rule i can be given as:

$$R_{obs}^{(i)} : \textbf{IF} \quad z_1 \quad is \ Q_{i1} \quad \textbf{AND} \quad z_2 \quad is \ Q_{i2} \quad \textbf{AND} \quad z_3 \quad is \ Q_{i3}$$
$$\textbf{THEN} \ u_1 \quad is \quad V_{obs\_i1} \quad \textbf{AND} \quad u_2 \quad is \quad \theta_{obs\_i2}, \quad i = 1, 2, \cdots, L$$

$$(6.13)$$

Table 6.2. Fuzzy rule base for obstacle avoidance

Rule No.	Antecedent Parts (IF clauses)			Consequence Parts (THEN parts)	
	$IR\_Group\_val(1)$	$IR\_Group\_val(2)$	$IR\_Group\_val(3)$	v_{obs} (in p.u.)	θ_{obs} (in degree)
1	FR	FR	FR	0.8	0
2	FR	FR	IM	0.8	90
3	FR	FR	NR	0.7	90
4	FR	IM	FR	0.5	135
5	FR	IM	IM	0.7	135
6	FR	IM	NR	0.6	150
7	FR	NR	FR	0.3	135
8	FR	NR	IM	0.5	135
9	FR	NR	NR	0.4	160
10	IM	FR	FR	0.9	90
11	IM	FR	IM	0.8	90
12	IM	FR	NR	0.6	90
13	IM	IM	FR	0.7	50
14	IM	IM	IM	0.3	90
15	IM	IM	NR	0.2	105
16	IM	NR	FR	0.5	35
17	IM	NR	IM	0.2	105
18	IM	NR	NR	0.1	150
19	NR	FR	FR	0.8	90
20	NR	FR	IM	0.7	90
21	NR	FR	NR	0.6	90
22	NR	IM	FR	0.5	40
23	NR	IM	IM	0.2	25
24	NR	IM	NR	0.1	90
25	NR	NR	FR	0.4	30
26	NR	NR	IM	0.1	15
27	NR	NR	NR	0	90

where

$$\mathbf{z} = [z_1, z_2, z_3]^T = [IR\_Group\_val(1), IR\_Group\_val(2), IR\_Group\_val(3)]^T,$$

$$\mathbf{u} = [u_1, u_2]^T = [v_{obs}, \theta_{obs}]^T,$$

$$Q_{i1} \in \{FR, IM, NR\}, Q_{i2} \in \{FR, IM, NR\}, Q_{i3} \in \{FR, IM, NR\}, V_{obs\_i1} \in \mathbf{v}_{obs} \quad \text{and}$$

$$\theta_{obs\_i2} \in \boldsymbol{\theta}_{obs}.$$

Here

$\mathbf{v}_{obs}$ = vector of singletons for the output linear velocity = $[v_{obs1}, v_{obs2}, \cdots, v_{obsL}]^T$ and

$\boldsymbol{\theta}_{obs}$ = vector of singletons for the output steering angle = $[\theta_{obs1}, \theta_{obs2}, \cdots, \theta_{obsL}]^T$.

L is the total number of rules for the fuzzy system and Table 4.2 shows the entire fuzzy rule base created for obstacle avoidance. Let K_{vel_obs} be the scaling gain for the linear velocity. Then the output crisp linear velocity command (v_{obs}) and the output steering angle command (θ_{obs}) are computed as:

$$v_{obs} = K_{vel\_obs} * \left(\frac{\sum_{i=1}^{L} v_{obsi} * \beta_i(\mathbf{z})}{\sum_{i=1}^{L} \beta_i(\mathbf{z})} \right) \tag{6.14}$$

$$\theta_{obs} = \frac{\sum_{i=1}^{L} \theta_{obsi} * \beta_i(\mathbf{z})}{\sum_{i=1}^{L} \beta_i(\mathbf{z})} \tag{6.15}$$

where $\beta_i(\mathbf{z})$ = firing degree of rule $i = \prod_{j=1}^{3} \mu_i(z_j)$.

6.5 Real-Life Performance Evaluation

Several experiments were conducted using the proposed system in real-life indoor environments. Four example case studies are reported here.

Case-Study I
In this study, the robot is commanded to follow a curved line. The width of the line is chosen smaller than the width of the robot. Figure 6.10 shows the sequence of images where the robot performs this commanded task. Figure 6.10(a) to Fig. 6.10(f) show a sequence of images when the robot is attempting to follow the middle of the line. For all the case studies, the navigation utilizes vision and IR

range sensing and hence the proximity sensors are turned off from the client side. Figure 6.11 shows the complete path of traversal in red colour. The ideal path of traversal for the robot is shown by the blue dotted line that goes through the middle of the line. It can be seen that the actual path traversed by the robot is in close agreement with the ideal path. At the corners, the actual path deviated a little more from the ideal path. This is understandable because a practical robot is expected to follow a smooth steering angle transition, when a fuzzy based navigation algorithm is employed.

Case-Study II

In this study, the robot is commanded to follow a bended line, take almost a U-turn when the path finishes, and trace the path back so that it can come back to its original starting position. Figure 6.12 shows the sequence of images where the robot performs this commanded task. This case study demonstrates the situation where the robot was commanded from the remote client to utilize the stored possible steering angle detour information to automatically turn, when the path in the forward direction finished, and attempt to come back on the line to trace its path back. Hence the robot did not stop when the path ended and both the *PixelCountLeft* and *PixelCountRight* computations produced zero values. Instead, it kept taking turns, in an iterative fashion, until it was able to retrace the path. Figure 6.12(a) to Fig. 6.12(f) show a sequence of images when the robot is in forward motion, attempting to follow the middle of the line.

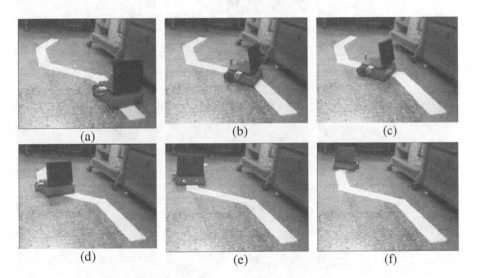

Fig. 6.10. Robot path traversed in case-study I

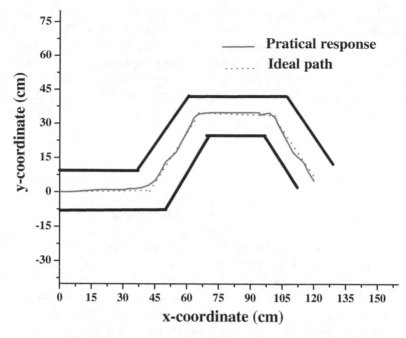

Fig. 6.11. The complete path of traversal for case study I

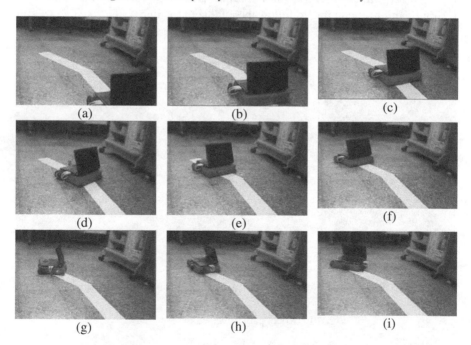

Fig. 6.12. Robot path traversed in case-study II

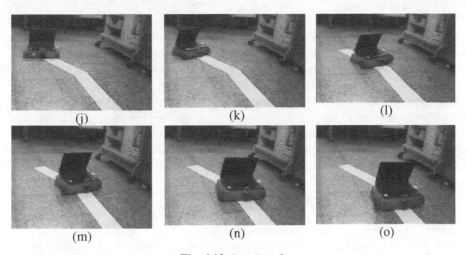

Fig. 6.12. (*continued*)

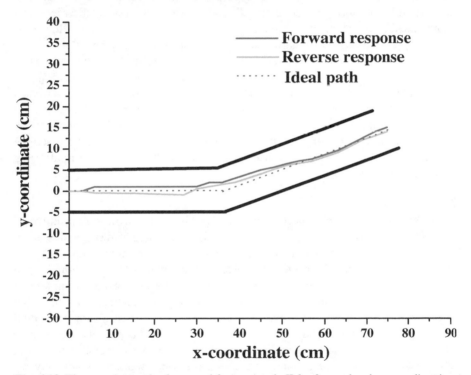

Fig. 6.13. The complete path of traversal for case study II for forward and reverse direction

Figure 6.12(g) to Fig. 6.12(k) show the sequence of images when the robot is performing the turning operation, in an iterative fashion, so that it can re-position itself on the line. Figure 6.12(l) to Fig. 6.12(o) show the next sequence of images when the robot was able to retrace that path and could come back following the line to its original starting point. A selection of "Search ON" option from the client end enabled the robot to attempt this retracing of path, even when the path disappeared from the field-of-view of its vision sensor. Figure 6.13 shows the forward path of traversal in red. The ideal path for the robot is shown by the blue dotted line that goes through the middle of the line. It can be seen that here also the actual path traversed by the robot is in close agreement with the ideal path and, at the corners, the actual path deviated a little more from the ideal path. It can also be seen that after taking almost a U-turn, the path traversed by the robot shown in green, had small deviations from the path traversed in the forward direction, shown in red. This shows a satisfactory performance for the robot, both while going up and then coming back.

Case-Study III
In this case study, the robot is commanded to follow the center of a path, which is of bigger width than the robot, as far as practicable, and there is an obstacle on the path which the robot needs to avoid. Figure 6.14 shows a sequence of images of how the robot performs its commanded task. In this experiment, the robot was commanded from the client end to navigate with "Search OFF" option. Hence the robot stopped after safely avoiding the obstacle and when it reached at the end of its path. As commanded, it did not attempt to re-localize itself on the path, when the path vanished from the field-of-view of the camera sensor. Figure 6.15 shows the complete path of traversal in red. This shows how the robot, at first, continued to travel along the middle of the road using vision sensing, and then, when it sensed the obstacle, took a left turn using IR based obstacle avoidance, safely avoided it by almost moving parallel to the obstacle, and then, when it crossed the obstacle, attempted to re-localize itself along the middle of the wide road, using vision sensing.

Case-Study IV
In this case study, the robot is commanded to perform a more difficult task, where the robot has to follow, on its way, two exactly perpendicular turns, and it is yet required to follow the middle of the line. In this case study the robot was commanded from the client end to navigate with "Search OFF" option. Figure 6.16 shows a sequence of images, when the robot performs this navigation task. It can be seen from these images that, in spite of these perpendicular turns, the robot was able to re-localize itself at the middle of the line, after each turn, and could follow the path commanded, in a satisfactory fashion. This can also be seen in Fig. 6.17 which shows the complete path of traversal. It can be seen that even after crossing the two perpendicular corners, the robot was able to quickly re-localize itself on the middle of the path and the deviation of the actual robot path from the ideal robot path is satisfactorily small. The deviations are a little more at the two perpendicular corners, which are again justifiable from the logic presented before.

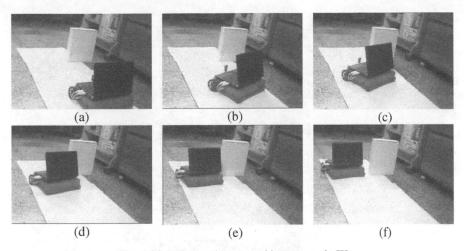

Fig. 6.14. Robot path traversed in case-study III

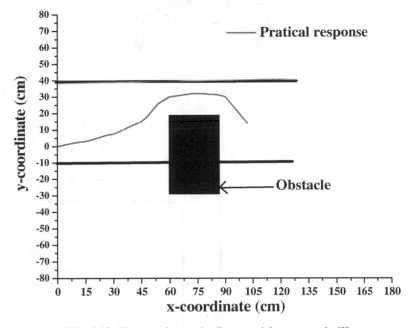

Fig. 6.15. The complete path of traversal for case study III

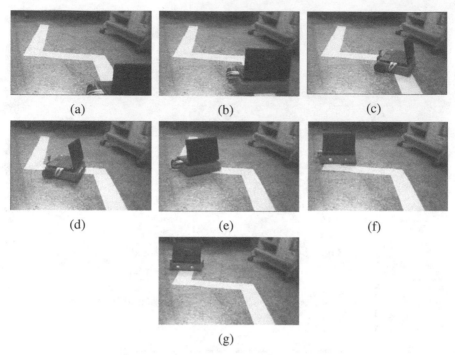

(a)   (b)   (c)

(d)   (e)   (f)

(g)

Fig. 6.16. Robot path traversed in case-study IV

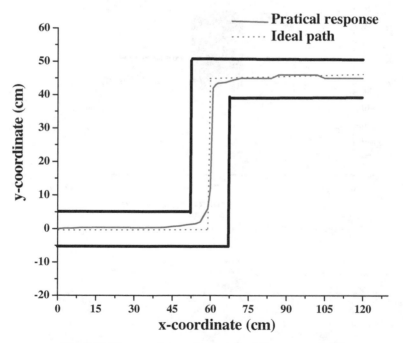

Fig. 6.17. The complete path of traversal for case study IV

6.6 Summary

This chapter discussed how a simple yet effective fuzzy based path/line tracking scheme can be implemented in indoor environments. The implementation is carried out for an indigenously developed mobile robot in our laboratory. The scheme employs fuzzy vision-based navigation, when the front of the robot is free of obstacles. When any obstacle is detected, a fuzzy IR-based obstacle avoidance scheme gets activated and the vision based scheme is deactivated. Once the obstacle is avoided, the IR-based scheme is deactivated and the vision based navigation takes over. The scheme always attempts to guide the robot along the middle of the path/line, whether the objective is to follow a narrow line or to travel along a wide path. The system has been implemented for navigation both in presence and in absence of obstacles and it has also been experimented successfully with intranet-connectivity functionality. This mobile robot path tracker system, implemented in conjunction with the indigenously developed robotic platform, has been experimented for several real-life indoor environments. Four such real-life case studies are discussed here to demonstrate the usefulness and effectiveness of the system developed.

Acknowledgement. The work described in this chapter was supported by University Grants Commission, India under Major Research Project Scheme (Grant No. 32-118/2006(SR)).

References

[1] Takeuchi, T., Nagai, Y., Enomoto, N.: Fuzzy control of a mobile robot for obstacle avoidance. Information Sciences 43, 231–248 (1988)

[2] Blochl, B.: Fuzzy Control in Real-Time for Vision Guided Autonomous Mobile Robots. In: Klement, E.-P., Slany, W. (eds.) FLAI 1993. LNCS, vol. 695, pp. 114–125. Springer, Heidelberg (1993)

[3] Yen, J., Pfluger, N.: A fuzzy logic based extension to Payton and Rosenblatt's command fusion method for mobile robot navigation. IEEE Transactions on Systems, Man, and Cybernatics 25(6), 971–978 (1995)

[4] Pan, J., Pack, D.J., Kosaka, A., Kak, A.C.: FUZZY-NAV: A vision-based robot navigation architecture using fuzzy inference for uncertainty. In: Proc. IEEE World Congress Neural Networks, vol. 2, pp. 602–607 (July 1995)

[5] Maeda, M., Shimakawa, M., Murakami, S.: Predictive fuzzy control of an autonomous mobile robot with forecast learning function. Fuzzy Sets and Systems 72, 51–60 (1995)

[6] Congdon, C., Huber, M., Kortenkamp, D., Konolige, K., Myers, K., Ruspini, E.H., Saffiotti, A.: CARMEL vs. Flakey: A comparison of two winners. Artificial Intelligence Magazine 14(1), 49–57 (1993)

[7] Goodridge, S.G., Luo, R.C.: Fuzzy behavior fusion for reactive control of an autonomous mobile robot: MARGE. In: Proc. IEEE International Conference on Robotics and Automation, San Diego, pp. 1622–1627 (1994)

[8]   Saffiotti, A.: The uses of fuzzy logic in autonomous robot navigation. Soft
       Computing 1, 180–197 (1997)
[9]   Howard, A., Seraji, H., Tunstel, E.: A rule-based fuzzy traversability index for
       mobile robot navigation. In: Proc. International Conference on Robotics and
       Automation, Korea, pp. 3067–3071 (May 2001)
[10]   Nirmal Singh, N.: Vision Based Autonomous Navigation of Mobile Robots. Ph.D.
       Thesis, Jadavpur University, Kolkata, India (2010)

Chapter 7
Simultaneous Localization and Mapping (SLAM) in Mobile Robots*

Abstract. This chapter first introduces the concept of SLAM for navigation of mobile robots and then describes the extended Kalman filter (EKF) based SLAM algorithms in detail. Next we consider a more complex scenario where this EKF based SLAM algorithm is implemented in presence of incorrect knowledge of sensor statistics and discuss how fuzzy or neuro-fuzzy supervision can help in improving the estimation performance in such situations. In this context, we also discuss how evolutionary optimization strategies can be employed to automatically learn the free parameters of such neuro-fuzzy supervisors.

7.1 Introduction

The simultaneous localization and mapping (SLAM) problem has attracted significant attention from the research communities of the autonomous vehicles and mobile robots in the past two decades. The SLAM problem, essentially, consists of estimating the unknown motion of a moving platform iteratively, in an unknown environment and, hence, determining the map of the environment consisting of features (also known as landmarks) and the absolute location of the moving platform on the basis of each other's information [1]. This is known as a very complex problem as there is always the possibility that both the vehicle's pose estimate and its associated map estimates become increasingly inaccurate in absence of any global position information [2]. This situation arises when a vehicle does not have access to a global positioning system (GPS). Hence the complexity of the SLAM problem is manifold and requires a solution in a high dimensional space due to the mutual dependence of vehicle pose and the map estimates [3].

---

* This chapter is based on:

1) "A neuro-fuzzy assisted extended Kalman filter-based approach for Simultaneous Localization and Mapping (SLAM) problems," by Amitava Chatterjee and Fumitoshi Matsuno, which appeared in *IEEE Transactions on Fuzzy Systems*, vol. 15, issue 5, pp. 984-997, October 2007. © 2007 IEEE and

2) Amitava Chatterjee, "Differential evolution tuned fuzzy supervisor adapted extended kalman filtering for SLAM problems in mobile robots," *Robotica*, vol. 27, issue 3, pp. 411-423, May 2009, reproduced with permission from Cambridge University Press.

One of the oldest and popular approaches to solve the SLAM problem employs Kalman filter based techniques. Until now extensive research works have been reported employing EKF to address several aspects of the SLAM problem [1], [4-12]. Several successful applications of SLAM algorithms have been developed for indoor applications [13, 14], outdoor applications [7], underwater applications [15], underground applications [16] etc. An EKF based approach estimates and stores the robot pose and the feature positions within the map of the environment in the form of a complete state-vector and the uncertainties in these estimates are stored in the form of error covariance matrices. These covariance matrices also include cross-correlation terms signifying cross-correlation among feature/landmark estimates. However, one of the well-known problems with the classical full EKF-based SLAM approach is that the computational burden becomes significantly high in the presence of a large number of features, because both the total state vector and the total covariance matrix become large in size. The later variations of researches on EKF based SLAMs have identified this problem as a key area and several improvements have so far been proposed [7, 9, 17-19]. Another key problem associated with EKF-based SLAM is the data association problem, which arises because several landmarks in the map may look similar. In those situations, different data association hypotheses can give rise to multiple, distinct looking maps and Gaussian distribution cannot be employed to represent such multi-modal distributions. This problem is usually solved by restricting the algorithm to associate the most likely data association, given the current robot map, on the basis of single measurement [1] or on the basis of multiple measurements [20]. The method of utilizing multiple measurements is a more robust method. Although several other data association algorithms have so far been developed, e.g. those in [21, 22], these algorithms have less significance as they cannot be implemented in real-time.

Some alternative approaches to solve SLAM problems have also been proposed which intend to implement some numerical algorithms, rather than employing the rigorous statistical methods as in EKF. Some of these schemes are based on the Bayesian approaches which can dispense with the important assumption in EKF (i.e. the uncertainties should be modeled by Gaussian distributions). Several such algorithms have been developed employing Sequential Monte Carlo (SMC) methods that employ the essence of particle filtering [2], [3], [23], [24]. Particle filtering technique can do away with a basic restriction of EKF algorithm that introduces an additional uncertainty by performing linearization of nonlinear models. However, in particle filtering based methods, it is expected that one should employ large number of particles so that it can contain a particle that can very closely resemble the true pose of the vehicle/robot at each sampling time instant [25]. How to develop an efficient SLAM algorithm, employing particle filtering with small enough number of particles, constitutes an important area of modern-day research. A significant leap in this direction is taken by the FastSLAM1.0 and FASTSLAM2.0 algorithms, which have successfully solved the issue of dimensionality for particle filter based SLAM problems [26]. Several other SLAM algorithms have also been successfully developed employing scan-matching technique where the map can efficiently be built by a graph of spatial relations amongst reference frames [7], [27].

It has been shown previously that the performance of an EKF process depends largely on the accuracy of the knowledge of process covariance matrix (**Q**) and measurement noise covariance matrix (**R**). An incorrect *a priori* knowledge of **Q** and **R** may lead to performance degradation [28] and it can even lead to practical divergence [29]. Hence adaptive estimation of these matrices becomes very important for online deployment. In [28], Mehra has reported a pioneering work on adaptive estimation of noise covariance matrices **Q** and **R** for Kalman filtering algorithm, based on correlation-innovations method, that can provide asymptotically normal, unbiased and consistent estimates of **Q** and **R** [35]. This algorithm is based on the assumption that noise statistics is stationary and the model under consideration is a time invariant one. Later several research works have been reported in the same direction, employing classical approaches, which have attempted adaptive estimation of **Q** and **R** [30-35]. In [30], a combination of an iterative algorithm and a stochastic approximation algorithm has been proposed to estimate **Q** and **R**. In [32] and [33], the problem domain has been expanded to allow time-variance in estimation of **Q** and **R**. A wonderful practical application of [28] has been reported in [34].

In the last ten years or so, there have also been several adaptive Kalman filtering algorithms proposed which employ fuzzy or neuro-fuzzy based techniques [36]-[39]. In [38], an input-output mapping problem, where output is corrupted by measurement noise, is solved by employing a neuro-fuzzy network to determine AR parameters of each operating point dependant ARMA model and then employing Kalman filter for the equivalent state-space representation of the system. In [36], fuzzy logic has been employed for simultaneous adaptive estimations of **Q** and **R** and in [37], fuzzy logic is employed to adapt the **R** matrix only, for a Kalman filter algorithm. In real world situations, it is quite perceptible that these information matrices, in the form of **Q** and **R**, may not be accurately known. Then the performance of the SLAM problem may get affected significantly.

The present chapter will first introduce the EKF-based stochastic SLAM algorithm in detail. Then the chapter will explore those situations for SLAM problems where the noise statistics information for the sensor is not known accurately. In those situations, we shall describe how neuro-fuzzy assisted EKF based SLAM algorithms can be effectively utilized [44, 45]. This will detail how a neuro-fuzzy model can be employed to assist the EKF-based SLAM algorithm to estimate **R** adaptively in each iteration. The chapter will also discuss how the free parameters of the neuro-fuzzy model can be learned using popular evolutionary optimization algorithms, for example, particle swarm optimization (PSO) [40] and differential evolution. The fuzzy adapted Kalman filter algorithms discussed in this chapter essentially implement a much complicated and sophisticated system compared to its predecessors mainly in two aspects:

i) For the SLAM problem, the situation is essentially very complex as the sizes of the state vector and hence the covariance matrix are time varying in nature. This is because, during the process of navigation, new landmarks are initialized in the state vector at different time instants (and, under some specific conditions, some existing landmarks may even be deleted) and hence these vector and matrix sizes will keep changing. The sizes of these matrices usually grow.

ii) The approaches discussed in this chapter uses a generalized method of learning the neuro-fuzzy model automatically. This is in stark contrast with previously developed systems which use carefully, manually chosen parameters for the fuzzy system(s) under consideration.

The chapter concludes with a detail, in-depth analysis of these SLAM algorithms where the results are presented for a variety of environmental situations i.e. with varying number of feature/landmark points and with several incorrectly known measurement noise statistics values.

7.2 Extended Kalman Filter (EKF) Based Stochastic SLAM Algorithm

A. *Hypotheses*

- The features under consideration are assumed to be 2-D point features
- The features are assumed to remain static i.e. they do not change their positions with time, in the map built
- There are uncertainties in control inputs, the steering angle command (*s*) and the velocity at which the rear wheel is driven (*w*), and these uncertainties are modeled using Gaussian distributions
- It is assumed that there is no uncertainty in the starting pose of the robot
- The incremental movement of the robot, between two successive sampling instants, is assumed to be linear in nature
- There are uncertainties in the range (*r*) and bearing (*θ*) measurements, and these uncertainties are modeled using Gaussian distributions
- The features are only characterized by their 2-D positions and no other characteristics, e.g. shape etc., is considered in this work

B. *The Algorithm*
An overview of the feature-map based SLAM employing EKF algorithm is presented now. An excellent description of the algorithm can also be obtained in [6], [7]. An EKF is employed for state estimation in those situations where the process is governed by nonlinear dynamics and/or involves nonlinear measurement relationships. The method employs linearization about the filter's estimated trajectory, which is continuously updated in accordance with the state estimates obtained from the measurements [43]. The state transition can be modeled by a nonlinear function $\mathbf{f}(\bullet)$ and the observation or measurement of the state can be modeled by a nonlinear function $\mathbf{h}(\bullet)$, given as:

$$\mathbf{x}_{k+1} = \mathbf{f}(\mathbf{x}_k, \mathbf{u}_k) + \mathbf{q}_k \tag{7.1}$$

and

$$\mathbf{z}_{k+1} = \mathbf{h}(\mathbf{x}_{k+1}) + \mathbf{r}_{k+1} \tag{7.2}$$

where $\mathbf{x}_k$ is the $(n \times 1)$ process state vector at sampling instant k, $\mathbf{z}_k$ is the $(m \times 1)$ measurement vector at sampling instant k and $\mathbf{u}_k$ is the control input. The random variables $\mathbf{q}_k$ and $\mathbf{r}_k$ represent Gaussian white process noise and measurement noise respectively and $\mathbf{P}_k$, $\mathbf{Q}_k$ and $\mathbf{R}_k$ represent the covariance matrices for $\mathbf{x}_k$, $\mathbf{q}_k$ and $\mathbf{r}_k$ respectively.

In case of the SLAM problem, the state vector $\mathbf{x}$ is composed of the vehicle states $\mathbf{x}_v$ and the landmarks' states $\mathbf{x}_m$. Hence the estimates of the total state vector $\mathbf{x}$, maintained in the form of its mean vector $\hat{\mathbf{x}}$ and the corresponding total error covariance matrix $\mathbf{P}$, is given as:

$$\hat{\mathbf{x}} = [\hat{\mathbf{x}}_v^T \quad \hat{\mathbf{x}}_m^T]^T \tag{7.3}$$

$$\mathbf{P} = \begin{bmatrix} \mathbf{P}_v & \mathbf{P}_{vm} \\ \mathbf{P}_{vm}^T & \mathbf{P}_m \end{bmatrix} \tag{7.4}$$

where $\hat{\mathbf{x}}_v$ = the mean estimate of the robot/vehicle states (represented by its pose),

$\mathbf{P}_v$ = error covariance matrix associated with $\hat{\mathbf{x}}_v$,

$\hat{\mathbf{x}}_m$ = mean estimate of the feature positions and

$\mathbf{P}_m$ = error covariance matrix associated with $\hat{\mathbf{x}}_m$.

The robot/vehicle pose is defined with respect to an arbitrary base Cartesian coordinate frame. The features or landmarks are considered to be 2-D point features. It is assumed that there are n such static, point features observed in the map. Then,

$$\hat{\mathbf{x}}_v = [\hat{x}_v \quad \hat{y}_v \quad \hat{\varphi}_v]^T \tag{7.5},$$

$$\mathbf{P}_v = \begin{bmatrix} \sigma_{x_v x_v}^2 & \sigma_{x_v y_v}^2 & \sigma_{x_v \varphi_v}^2 \\ \sigma_{x_v y_v}^2 & \sigma_{y_v y_v}^2 & \sigma_{y_v \varphi_v}^2 \\ \sigma_{x_v \varphi_v}^2 & \sigma_{y_v \varphi_v}^2 & \sigma_{\varphi_v \varphi_v}^2 \end{bmatrix} \tag{7.6},$$

$$\hat{\mathbf{x}}_m = [\hat{x}_1 \quad \hat{y}_1 \quad \cdots\cdots \quad \hat{x}_n \quad \hat{y}_n]^T \tag{7.7}$$

and

$$\mathbf{P}_m = \begin{bmatrix} \sigma_{x_1 x_1}^2 & \sigma_{x_1 y_1}^2 & \cdots & \sigma_{x_1 x_n}^2 & \sigma_{x_1 y_n}^2 \\ \sigma_{x_1 y_1}^2 & \sigma_{y_1 y_1}^2 & \cdots & \sigma_{y_1 x_n}^2 & \sigma_{y_1 y_n}^2 \\ \vdots & \vdots & \ddots & \vdots & \vdots \\ \sigma_{x_1 x_n}^2 & \sigma_{y_1 x_n}^2 & \cdots & \sigma_{x_n x_n}^2 & \sigma_{x_n y_n}^2 \\ \sigma_{x_1 y_n}^2 & \sigma_{y_1 y_n}^2 & \cdots & \sigma_{x_n y_n}^2 & \sigma_{y_n y_n}^2 \end{bmatrix} \tag{7.8}$$

The map is defined in terms of the position estimates of these static features and $\mathbf{P}_{vm}$ in (7.4) maintains the robot-map correlation. The off-diagonal elements of $\mathbf{P}_m$

signify the cross-correlation and hence interdependence of information among the features themselves. The system is initialized assuming that there is no observed feature as yet, the base Cartesian coordinate frame is aligned with the robot's starting pose and there is no uncertainty in the starting pose of the robot. Mathematically speaking, $\hat{\mathbf{x}} = \hat{\mathbf{x}}_v = \mathbf{0}$ and $\mathbf{P} = \mathbf{P}_v = \mathbf{0}$.

As the robot starts moving, $\hat{\mathbf{x}}_v$ and $\mathbf{P}_v$ become non-zero values. In subsequent iterations, when the first observation is carried out, new features are expected to be initialized and $\hat{\mathbf{x}}_m$ and $\mathbf{P}_m$ appear for the first time. This increases the size of $\hat{\mathbf{x}}$ and $\mathbf{P}$ and the entries of $\hat{\mathbf{x}}$ vector and $\mathbf{P}$ matrix are re-calculated. This process is continued iteratively.

i) Time Update ("Predict") Step

Here, it is assumed that the control input vector $\mathbf{u}$, under the influence of which the robot moves, is constituted of two control inputs, the steering angle command (s) and the velocity at which the rear wheel is driven (w). Hence, $\mathbf{u} = [w \ s]^T$. So the state estimates can be obtained by employing wheel encoder odometry and the robot kinematic model. The control inputs w and s must be considered with their uncertainties involved (e.g. uncertainties due to wheel slippage, incorrect calibration of vehicle controller) and these are modeled as Gaussian variations in w and s from their nominal values. Hence, the prediction step calculates the projections of the state estimates and the error covariance estimates from sampling instant k to $(k+1)$, given as:

$$\hat{\mathbf{x}}_{k+1}^- = \mathbf{f}(\hat{\mathbf{x}}_k, \hat{\mathbf{u}}_k) = \begin{bmatrix} \hat{\mathbf{x}}_{v_{k+1}}^- \\ \hat{\mathbf{x}}_m \end{bmatrix} = \begin{bmatrix} \mathbf{f}_v(\hat{\mathbf{x}}_{v_k}, \hat{\mathbf{u}}_k) \\ \hat{\mathbf{x}}_m \end{bmatrix} \qquad (7.9)$$

$$\mathbf{P}_{k+1}^- = \begin{bmatrix} \nabla \mathbf{f}_{\mathbf{x}_{v_k}} \mathbf{P}_{v_k} \nabla \mathbf{f}_{\mathbf{x}_{v_k}}^T + \nabla \mathbf{f}_{\mathbf{u}_k} \mathbf{U}_k \nabla \mathbf{f}_{u_k}^T & \nabla \mathbf{f}_{\mathbf{x}_{v_k}} \mathbf{P}_{vm_k} \\ (\nabla \mathbf{f}_{\mathbf{x}_{v_k}} \mathbf{P}_{vm_k})^T & \mathbf{P}_m \end{bmatrix} \qquad (7.10)$$

where $\mathbf{f}_v$ estimates the robot pose on the basis of the motion model and the control inputs. Based on the odometric equation of the mobile robot under consideration here, which assumes that the incremental movement of the robot is linear in nature, $\mathbf{f}_v$ can be represented as [42]:

$$\hat{\mathbf{x}}_{v_{k+1}}^- = \begin{bmatrix} \hat{x}_{v_{k+1}}^- \\ \hat{y}_{v_{k+1}}^- \\ \hat{\varphi}_{v_{k+1}}^- \end{bmatrix} = \mathbf{f}_v(\hat{\mathbf{x}}_{v_k}, \hat{\mathbf{u}}_k) = \begin{bmatrix} \hat{x}_{v_k} + w_k * \Delta t * \cos(s_k + \hat{\varphi}_{v_k}) \\ \hat{y}_{v_k} + w_k * \Delta t * \sin(s_k + \hat{\varphi}_{v_k}) \\ \hat{\varphi}_{v_k} + w_k * \Delta t * \dfrac{\sin(s_k)}{WB} \end{bmatrix} \qquad (7.11)$$

where, *WB* represents the wheelbase of the robot and Δt is the sampling time. The Jacobians and $\mathbf{U}_k$, the covariance matrix of $\mathbf{u}$ are given as:

$$\nabla \mathbf{f}_{\mathbf{x}_{v_k}} = \frac{\partial \mathbf{f}_{v_k}}{\partial \mathbf{x}_{v_k}}\bigg|_{(\hat{\mathbf{x}}_{v_k}, \hat{\mathbf{u}}_k)} \quad (7.12), \qquad \nabla \mathbf{f}_{\mathbf{u}_k} = \frac{\partial \mathbf{f}_{v_k}}{\partial \mathbf{u}_k}\bigg|_{(\hat{\mathbf{x}}_{v_k}, \hat{\mathbf{u}}_k)} \quad (7.13), \quad \mathbf{U} = \begin{bmatrix} \sigma_v^2 & 0 \\ 0 & \sigma_s^2 \end{bmatrix}$$

$$(7.14)$$

Here, $\hat{\mathbf{x}}_m$ and $\mathbf{P}_m$ in (7.9) and (7.10) remain constant with time, as the features are assumed to remain stationary with time.

ii) Measurement Update ("Correct") Step
Let us assume that we observe a feature, which already exists in the feature map, whose position is denoted by that of the *i*th feature i.e. $(\hat{x}_i, \hat{y}_i)$. For the system under consideration [7], [42], it is assumed that the feature observation is carried out using 2-D scanning range laser (SICK PLS), a range-bearing sensor, which nowadays is very popular in mobile robot navigation, for distance measurement. It is assumed that the laser range scanner is mounted on the front bumper of the vehicle and the laser returns a 180° planar sweep of range measurements in 0.5° intervals. The range resolution of such a popular sensor is usually about ±50 mm. In this context, it should be mentioned that the vehicle is also assumed to be equipped with wheel and steering encoders. The distance measured, in polar form, gives the relative distance between each feature and the scanner (and hence the vehicle). Let this feature be measured in terms of its range (*r*) and bearing (*θ*) relative to the observer, given as:

$$\mathbf{z} = [r \ \ \theta]^T \tag{7.15}$$

The uncertainties in these observations are again modeled by Gaussian variations and let $\mathbf{R}$ be the corresponding observation/measurement noise covariance matrix given as:

$$\mathbf{R} = \begin{bmatrix} \sigma_r^2 & 0 \\ 0 & \sigma_\theta^2 \end{bmatrix} \tag{7.16}$$

where we assume that there is no cross-correlation between the range and bearing measurements. In the context of the map, the measurements can be given as:

$$\hat{\mathbf{z}}_{i_k} = \mathbf{h}_i(\hat{\mathbf{x}}_k) = \begin{bmatrix} \sqrt{(\hat{x}_i - \hat{x}_{v_k})^2 + (\hat{y}_i - \hat{y}_{v_k})^2} \\ \arctan(\dfrac{\hat{y}_i - \hat{y}_{v_k}}{\hat{x}_i - \hat{x}_{v_k}}) - \hat{\varphi}_{v_k} \end{bmatrix} \tag{7.17}$$

Now the Kalman gain $\mathbf{W}_i$ can be calculated assuming that there is correct landmark association between $\mathbf{z}$ and $(\hat{x}_i, \hat{y}_i)$ and the following computations can be resorted to:

$$v_{i_{k+1}} = \mathbf{z}_{k+1} - \mathbf{h}_i(\hat{\mathbf{x}}_{k+1}^-) \qquad (7.18)$$

$$\mathbf{S}_{i_{k+1}} = \nabla\mathbf{h}_{\mathbf{x}_{k+1}} \mathbf{P}_{k+1}^- \nabla\mathbf{h}_{\mathbf{x}_{k+1}}^T + \mathbf{R}_k \qquad (7.19)$$

$$\mathbf{W}_{i_{k+1}} = \mathbf{P}_{k+1}^- \nabla\mathbf{h}_{\mathbf{x}_{k+1}}^T \mathbf{S}_{i_{k+1}}^{-1} \qquad (7.20)$$

where v_i denotes the innovation of the observation for this ith landmark and $\mathbf{S}_i$ the associated innovation covariance matrix. The Jacobian $\nabla\mathbf{h}_{\mathbf{x}_{k+1}}$ is given as:

$$\nabla\mathbf{h}_{\mathbf{x}_{k+1}} = \frac{\partial\mathbf{h}_i}{\partial\mathbf{x}_k}\bigg|_{\hat{\mathbf{x}}_{k+1}^-} \qquad (7.21)$$

Hence, the *a posterior* augmented state estimate and the corresponding covariance matrix are updated as:

$$\hat{\mathbf{x}}_{k+1}^+ = \hat{\mathbf{x}}_{k+1}^- + \mathbf{W}_{i_{k+1}} v_{i_{k+1}} \qquad (7.22)$$

$$\mathbf{P}_{k+1}^+ = \mathbf{P}_{k+1}^- - \mathbf{W}_{i_{k+1}} \mathbf{S}_{i_{k+1}} \mathbf{W}_{i_{k+1}}^T \qquad (7.23)$$

Here it should be remembered that in addition to the process and measurement uncertainties, there is an additional uncertainty due to linearization involved in the formulation of an EKF. The "time update" and "measurement update" equations are obtained by employing linearization of nonlinear functions $\mathbf{f}(\bullet)$ and $\mathbf{h}(\bullet)$ about the point of the state mean. This linearization is obtained by employing a Taylor series like expansion and neglecting all terms which are of higher order than the first order term in the series. This manner of approximating a nonlinear system by a first order derivative introduces this additional source of uncertainty in EKF algorithm. In fact, for highly nonlinear functions, these linearized transformations cannot sufficiently accurately approximate correct covariance transformations and this may lead to highly inconsistent uncertainty estimate. Under those situations unscented transform may provide more accurate results.

iii) Initialization of a new feature and deletion of an old feature
During this iterative procedure of performing prediction and update steps recursively, it is very likely that observations of new features are made time to time. Then these new features should be initialized into the system by incorporating their 2-D position coordinates in the augmented state vector and accordingly modifying the covariance matrix. These features, identified by the

LRS, may correspond to points, lines, corners, edges etc. In this work, we have considered that the features are point like features, each representing a unique distinct point in the two-dimensional map of the environment. Resorting to the mathematical computations as shown in [7], these new $\hat{\mathbf{x}}_k^+$ and $\mathbf{P}_k^+$ can be calculated as:

$$\hat{\mathbf{x}}_k^+ = \begin{bmatrix} \hat{\mathbf{x}}_k \\ \mathbf{f}_f(\hat{\mathbf{x}}_{v_k}, \mathbf{z}_k) \end{bmatrix} \tag{7.24}$$

$$\mathbf{P}_k^+ = \begin{bmatrix} \mathbf{P}_{v_k} & \mathbf{P}_{vm_k} & \mathbf{P}_{v_k} \nabla \mathbf{f}_{\mathbf{f}_{v_k}}^T \\ \mathbf{P}_{vm_k}^T & \mathbf{P}_m & \mathbf{P}_{vm_k}^T \nabla \mathbf{f}_{\mathbf{f}_{v_k}}^T \\ \nabla \mathbf{f}_{\mathbf{f}_{v_k}} \mathbf{P}_{v_k} & \nabla \mathbf{f}_{\mathbf{f}_{v_k}} \mathbf{P}_{vm_k} & \nabla \mathbf{f}_{\mathbf{f}_{v_k}} \mathbf{P}_{v_k} \nabla \mathbf{f}_{\mathbf{f}_{v_k}}^T + \nabla \mathbf{f}_{\mathbf{f}_{z_k}} \mathbf{R}_k \nabla \mathbf{f}_{\mathbf{f}_{z_k}}^T \end{bmatrix} \tag{7.25}$$

Here $\mathbf{f}_f(\hat{\mathbf{x}}_v, \mathbf{z})$ is employed to convert the polar observation $\mathbf{z}$ to the base Cartesian coordinate frame. The Jacobians are calculated as:

$$\nabla \mathbf{f}_{\mathbf{f}_{v_k}} = \frac{\partial \mathbf{f}_f}{\partial \mathbf{x}_{v_k}} \bigg|_{(\hat{\mathbf{x}}_{v_k}, \mathbf{z}_k)} \quad , \quad \nabla \mathbf{f}_{\mathbf{f}_{z_k}} = \frac{\partial \mathbf{f}_f}{\partial \mathbf{z}_k} \bigg|_{(\hat{\mathbf{x}}_{v_k}, \mathbf{z}_k)} \tag{7.26}$$

The deletion of unreliable features is a relatively simple matter. We only need to delete the relevant row entries from the state vector and the relevant row and column entries from the covariance matrix.

Now, it is quite common that when an observation step is carried out, there will be multiple number of landmarks visible at the same time and hence, several independent observations will be carried out. In our system, we have assumed that a batch of such observations is available at once (i.e. $\mathbf{z} = [r_1, \theta_1, \cdots r_n, \theta_n]^T$) and updates are carried out in batches. This is in conformation with the arguments placed in [7] which indicate that an EKF algorithm tends to perform better update steps for SLAM algorithms, if the innovation vector v consists of multiple observations simultaneously. Hence, in the context of this batch mode of observation and update procedure, the corresponding SLAM algorithm is based on composite v, $\mathbf{S}$ and $\mathbf{W}$ vectors/matrices and the sizes of these vectors/matrices keep changing with time because at any instant of observation, the total number of visible landmarks keep changing.

7.3   Neuro-fuzzy Assistance for EKF Based SLAM

Most of the works reported in the area of adaptive Kalman filters have so far concentrated on utilizing new statistical information from innovation sequence to correct the estimation of the states. Our approach for adapting the EKF is based on the innovation adaptive estimation (IAE) approach, which was originally proposed in [28] and later utilized in combination with fuzzy logic in [37]. The basic concept relies on determining the discrepancy between a new measurement $\mathbf{z}_k$ and its corresponding predicted estimation $\hat{\mathbf{z}}_k$, at any arbitrary kth instant, and utilizing this new information to correct the estimations/predictions already made. The adaptation strategy is based on the objective of reducing mismatch between the theoretical covariance of the innovation sequences ($\mathbf{S}_k$) and the corresponding actual covariance of the innovation sequences ($\hat{\mathbf{C}}_{Innk}$). In our SLAM algorithm, $\mathbf{S}_k$ is calculated using (7.19) where the right hand side of the equation is made consistent with the concept of batch mode of observation and update. $\hat{\mathbf{C}}_{Innk}$ can be calculated as:

$$\hat{\mathbf{C}}_{Innk} = v_k \, v_k^T \qquad (7.27)$$

where v_k denotes the augmented innovation sequence, made consistent with the batch mode. According to [37], this covariance should be calculated on the basis of a moving average of $v_k \, v_k^T$ over an appropriate moving estimation window of size M. However, for the SLAM problem, the size of the augmented v_k keeps changing from time to time. This is because it is dependent on the number of landmarks observed in any given *observation and update step*, which were all observed at least once before. Hence we employ (7.27) to calculate $\hat{\mathbf{C}}_{Innk}$ rather than using a moving average. Therefore, the mismatch at the kth instant, is given as:

$$\Delta\hat{\mathbf{C}}_{Innk} = \hat{\mathbf{C}}_{Innk} - \mathbf{S}_k \qquad (7.28)$$

Our objective is to minimize this mismatch employing fuzzy logic. This is carried out, by employing a one-input-one-output neuro-fuzzy system for each diagonal element of the $\Delta\hat{\mathbf{C}}_{Innk}$ matrix. These fuzzy rules are employed to adapt the $\mathbf{R}$ matrix, so that the sensor statistics is adapted for subsequent reduction in mismatch $\Delta\hat{\mathbf{C}}_{Innk}$. The complete EKF-based SLAM algorithm, employing the neuro-fuzzy assistance, is presented in algo. 7.1. The system is designed with a sampling time of 25 msec. between successive control input signals.

1. *IF* All waypoints are traversed, *THEN* Stop *ENDIF*.
2. Compute distance of the robot from the current waypoint.
    IF (distance < minimum distance allowed from any waypoint),
        THEN switch to next waypoint as the current waypoint *ENDIF*.
3. Compute change in steering angle (Δs) to point towards the current waypoint and then, new value of steering angle (s) (satisfying the constraints of max. rate of steering change (Δs_{max}) and max. steering angle (s_{max})).
4. Move the robot and determine its actual pose.
5. Perform EKF prediction step, in accordance with (7.7) to (7.10).
6. *IF* (*Time_for_Observation* is TRUE), *THEN* go to step 7. *ELSE* go to step 1. *ENDIF*.
7. Determine the set of visible landmarks from the current actual robot position. Compute actual range-bearing observation for each of them. Separate those observations based on already observed landmarks and newly observed landmarks (if any).
8. Predict range-bearing observations, for already observed landmarks in step 7, on the basis of augmented total state vector, predicted in step 5.
9. Compute augmented innovation sequence (ν) for already observed landmarks, on the basis of actual and predicted observations, employing (7.14), adapted for batch-mode situations.
10. Compute corresponding augmented measurement noise covariance matrix **R** (utilizing the original [2×2] **R** matrix) and augmented linearized observation model **h,** adapted for batch-mode situations.
11. Compute augmented **S**, on the basis of the augmented **R** and **h** and employing (7.15), adapted for batch-mode situations.
12. Update the *a posterior* state estimate vector and error covariance matrix, according to (7.18) and (7.19).
13. Compute $\hat{\mathbf{C}}_{Innk}$ and $\Delta\hat{\mathbf{C}}_{Innk}$, according to (7.23) and (7.24) respectively, and determine the size of $\Delta\hat{\mathbf{C}}_{Innk}$, i.e. $[\Delta\hat{\mathbf{C}}_{rows}, \Delta\hat{\mathbf{C}}_{cols}]$.
14. Determine the absolute maximum value of mismatch among the range observations ($\Delta\hat{\mathbf{C}}_{Innk\_range\_mismatch}$) and the bearing observations ($\Delta\hat{\mathbf{C}}_{Innk\_bearing\_mismatch}$) separately from the corresponding diagonal entries of the $\Delta\hat{\mathbf{C}}_{Innk}$ matrix.
15. *FOR* $j = 1$ to $\Delta\hat{\mathbf{C}}_{rows}$,

    Normalize the corresponding diagonal entry $\Delta\hat{\mathbf{C}}_{Innk}(j, j)$ by the appropriate

    $\Delta\hat{\mathbf{C}}_{Innk\_range\_mismatch}$ or $\Delta\hat{\mathbf{C}}_{Innk\_bearing\_mismatch}$.

Determine the corresponding $\Delta\mathbf{R}(j, j)$ output from the NFS, with the normalized $\Delta\hat{\mathbf{C}}_{Innk}(j, j)$ input to it.

ENDFOR

16. Determine $\Delta\sigma_r^2$ as a mean of those $\Delta\mathbf{R}(j, j)$ entries, which correspond to range measurements.

17. Determine $\Delta\sigma_\theta^2$ as a mean of those $\Delta\mathbf{R}(j, j)$ entries, which correspond to bearing measurements.

18. Adapt the original 2×2 $\mathbf{R}$ matrix as: $\mathbf{R}_k = \mathbf{R}_{k-1} + \begin{bmatrix} \Delta\sigma_r^2 & 0 \\ 0 & \Delta\sigma_\theta^2 \end{bmatrix}$.

19. ***IF*** (new feature(s) observed in step 7),
    THEN augment state vector and error covariance matrix, according to (7.20),
       (7.21) and (7.22).
    ENDIF
20. Go to step 1.

Algo. 7.1. The neuro-fuzzy assisted EKF based SLAM algorithm

From algo. 7.1, it can be seen that each Neuro-Fuzzy System (NFS) employs a nonlinear mapping of the form: $\Delta\mathbf{R}(j, j) = f_{NFS}(\Delta\hat{\mathbf{C}}_{Innk}(j, j))$ where $\Delta\mathbf{R}(j, j)$ corresponds to an adaptation recommended for the corresponding diagonal element of the augmented measurement noise covariance matrix $\mathbf{R}$ matrix, computed according to the batch-mode situation. This augmented matrix is calculated each time an iteration enters into the *observe and update step* and its size is determined on the basis of the total landmarks visible in the *observe step*. To make it consistent with the batch of observed landmarks that were already visited at least once earlier, the size of this augmented $\mathbf{R}$ is $[2z_f \times 2z_f]$ where z_f is the number of landmarks observed in that iteration, which were also observed earlier. This augmented $\mathbf{R}$ is formed utilizing the original $[2 \times 2]$ $\mathbf{R}$ matrix and this is formulated as:

$$\text{augmented } \mathbf{R} = \begin{bmatrix} \sigma_r^2 & 0 & 0 & 0 & \cdots & 0 & 0 \\ 0 & \sigma_\theta^2 & \cdots & \cdots & \cdots & 0 & 0 \\ \vdots & \vdots & \sigma_r^2 & 0 & \cdots & \vdots & \vdots \\ \vdots & \vdots & 0 & \sigma_\theta^2 & \cdots & \vdots & \vdots \\ \vdots & \vdots & \cdots & \cdots & \ddots & \vdots & \vdots \\ 0 & 0 & \cdots & \cdots & \cdots & \sigma_r^2 & 0 \\ 0 & 0 & \cdots & \cdots & \cdots & 0 & \sigma_\theta^2 \end{bmatrix} \quad (7.29)$$

Here, σ_r^2 and σ_θ^2 correspond to the sensor statistics computed for that iteration. It can be seen that the augmented **R** matrix comprises of diagonal elements only and all the off-diagonal elements are considered to be zero. This is in conformation with our assumptions presented beforehand, in section 7.2, that the range and the bearing measurements are independent of each other and there is no cross-correlation between these measurements. The size of this augmented **R** matrix keeps changing in different iterations, as the number of already visited landmarks observed again in a given iteration keeps varying from iteration to iteration. The size of this augmented **R** is consistent with that of the $\hat{\mathbf{C}}_{Innk}$ and hence, $\Delta\hat{\mathbf{C}}_{Innk}$.

With the idea of implementing the same NFS for each and every diagonal element of the augmented **R** matrix, we employ normalized input for each NFS. The NFS practically employs three fuzzy *IF-THEN* rules of the form:

IF $\Delta\hat{\mathbf{C}}_{Innk}(j, j)$ is N   *THEN* $\Delta\mathbf{R}(j, j) = w_1$,

IF $\Delta\hat{\mathbf{C}}_{Innk}(j, j)$ is Z   *THEN* $\Delta\mathbf{R}(j, j) = w_2$ and

IF $\Delta\hat{\mathbf{C}}_{Innk}(j, j)$ is P   *THEN* $\Delta\mathbf{R}(j, j) = w_3$.

w_1, w_2 and w_3 indicate the amount of fuzzy adaptation recommended in form of a diagonal element of the $\Delta\mathbf{R}$ matrix, depending on the nature of the fuzzified mismatch in the corresponding diagonal element of the $\Delta\hat{\mathbf{C}}_{Innk}$ matrix. However, the order of mismatch may be different for range and bearing observations and this may depend on how poorly (or accurately) the sensor statistics for range and bearing observations are individually known. Hence we employ normalized inputs corresponding to range and bearing observations separately, on the basis of appropriate computations of $\Delta\hat{\mathbf{C}}_{Innk\_range\_mismatch}$ and $\Delta\hat{\mathbf{C}}_{Innk\_bearing\_mismatch}$, as given in algo. 7.1. Then with these normalized inputs, the NFS enables us to compute $\Delta\mathbf{R}(j, j)$ for each diagonal entry. Finally we compute the adaptations i.e. $\Delta\sigma_r^2$ and $\Delta\sigma_\theta^2$ required for the original [2 × 2] **R** matrix on the basis of appropriate means, separately computed from the arrays of $\Delta\mathbf{R}(j, j)$ entries for range and bearing observations. This adapted original [2 × 2] **R** matrix is kept ready for the next appropriate iteration, when EKF will enter the *observation and update step*, and will be utilized for subsequent formation of augmented **R** matrix and so on. Then, each *observation and update step* is concluded by augmenting the state vector and the corresponding covariance matrix, by employing (7.24)-(7.26), if there are new feature(s) observed during this *observation step*.

7.4   The Neuro-fuzzy Architecture and Its Training Methodology Employing Particle Swarm Optimization (PSO)

7.4.1 Architecture of the Neuro-fuzzy Model

The neuro-fuzzy model has been developed as a one-input-one-output system. The four-layer architecture is shown in Fig. 7.1. Let u_i^l and O_i^l respectively denote the input to and output from the ith node of the lth layer.

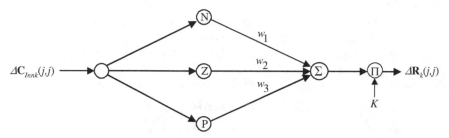

Fig. 7.1. Four-layer architecture of the proposed neuro-fuzzy system. (Reproduced from [44] with permission from the IEEE. ©2007 IEEE.).

1) Layer 1: Input Layer
This layer comprises a single node, signifying the single input variable. The input-output relation of this node is:

$$O^1 = u^1 = \Delta\hat{\mathbf{C}}_{Innk}(j, j) \tag{7.30}$$

2) Layer 2: Membership Function Layer
Here, the input variable is fuzzified employing three Membership Functions (MFs), negative (N), zero (Z) and positive (P). Figure 7.2 shows these MFs where N_v and N_b respectively denote the right vertex and right base points of the MF N, Z_{bl}, Z_{vl}, Z_{vr} and Z_{br} respectively denote the left base, left vertex, right vertex and right base points of the MF Z and P_b and P_v respectively denote the left base and left vertex points of the MF P. The output of the ith MF is given as:

$$O_i^2 = \mu_i(u^1) = \mu_i(\Delta\hat{\mathbf{C}}_{Innk}(j, j)) \tag{7.31}$$

3) Layer 3: Defuzzification layer
This layer performs defuzzification where the defuzzified output is calculated as an weighted average of all its inputs. Hence the output from the solitary node in this layer can be calculated as:

$$O^3 = \frac{\sum_{i=1}^{3} O_i^2 * w_i}{\sum_{i=1}^{3} O_i^2} = \frac{\sum_{i=1}^{3} \mu_i(u^1) * w_i}{\sum_{i=1}^{3} \mu_i(u^1)} \tag{7.32}$$

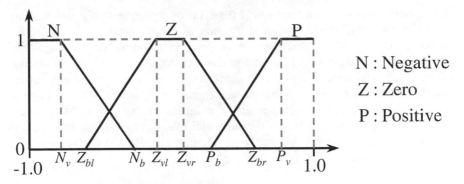

N : Negative

Z : Zero

P : Positive

Fig. 7.2. Membership functions employed in Fig. 7.1. (Reproduced from [44] with permission from the IEEE. ©2007 IEEE.).

4) Layer 4: Output Layer
This layer performs a suitable scaling for the defuzzified output. The input-output relationship of the node in this layer is given as:

$$O^4 = K * u^4 = K * O^3 \tag{7.33}$$

7.4.2 Training the Neuro-fuzzy Model Employing PSO

This neuro-fuzzy model is trained to determine the suitable free parameters of the system i.e. the parameters of the MFs, the output consequence singletons and the output gain. However, the training cannot be accomplished in the conventional supervised mode, as the exact desired output, for a given input, is not quantitatively known. Hence, normal backpropagation kind of training methodology cannot be resorted to and it is suitable to apply stochastic global optimization algorithms for such systems in an unsupervised manner. There are several such candidate algorithms available now. In this section we describe how PSO can be suitably employed for this purpose. PSO is a relatively new algorithm [40], [41], that is based on the swarm behaviors of birds or fishes. The training of the neuro-fuzzy system is accomplished as a high-dimensional metaheuristic optimization problem, where the objective is to optimize a fitness function $f_{fit}(x_1, x_2, \cdots x_n)$ on the basis of the values of the variables $x_1, x_2, \cdots x_n$.

In a PSO problem, several such candidate solutions of $x_1, x_2, \cdots x_n$ are created in a multi-dimensional space (called "particles") and the suitability of each of them is evaluated in each iteration. For the problem under consideration here, each such potential "particle" is formed as a 12-dimensional vector $\mathbf{x} = [x_1 \ x_2 \cdots x_{12}]^T$, as shown in Fig. 7.3. Each "particle" i is characterized by the vectors denoting its position ($\mathbf{x}_i$) and its velocity ($\mathbf{v}_i$) at the current time step. In order to pursue the optimum of the fitness function (f_{fit}), velocity $\mathbf{v}_i$ and hence position $\mathbf{x}_i$ of each particle is adjusted in each time step. The updated velocity in each time step $\mathbf{v}_{inew}$ is a function of three major components: the old velocity vector of the same particle ($\mathbf{v}_{iold}$), difference of the ith particle's best position found so far (called $\mathbf{p}_i$) and the current position ($\mathbf{x}_i$) (called the "cognitive" component) and difference of the best position of any particle within the context of the topological neighborhood of ith particle found so far (called $\mathbf{p}_g$) and current position of the ith particle ($\mathbf{x}_i$) (called the "social" component) [40, 41]. Each of the last two components is stochastically weighted so that the updating in the velocity of each particle will cause enough oscillations, allowing each particle to search for a better pattern within the problem space. Hence, the velocity and position update relations, in the dth dimension, are given as:

$$v_{idnew} = \quad v_{idold} + \varphi_i(p_{id} - x_{id}) + \varphi_g(p_{gd} - x_{gd})$$

$$\textbf{IF } (v_{idnew} > v_{dmax}) \textbf{ THEN } v_{idnew} = v_{dmax} \textbf{ ENDIF}$$

$$\textbf{IF } (v_{idnew} < -v_{dmax}) \textbf{ THEN } v_{idnew} = -v_{dmax} \textbf{ ENDIF}$$

$$x_{idnew} = x_{idold} + v_{idnew}$$

$$v_{idold} = v_{idnew}$$

$$x_{idold} = x_{idnew} \tag{7.34}$$

φ_i and φ_g are responsible for introducing stochastic weighting and they are given as $\varphi_i = c_i * rand_1(\ )$ and $\varphi_g = c_g * rand_2(\ )$. $rand_1(\ )$ and $rand_2(\ )$ are two random functions in [0, 1] and c_i and c_g are positive constants. A popular choice for c_i and c_g is $c_i = c_g = 2$. This traditional PSO model shows quick, aggressive convergence during the early phase but often encounters problem in fine tuning the search to determine the supreme solution. Hence, in our algorithm we have employed an improved version of this PSO algorithm that utilizes a judicious mix of aggressive, coarse updating during early iterations and fine updating during later iterations [40]. Hence the velocity update rule is given as

$$v_{idnew} = w_{iter}(v_{idold}) + \varphi_i(p_{id} - x_{id}) + \varphi_g(p_{gd} - x_{id}) \tag{7.35}$$

with the position update rule remaining unchanged as given before. w is called the inertia weight which is initially kept high and then gradually decreased over the iterations so that it can initially introduce coarse adjustment in velocity updating and gradually fine changes in velocity updating takes over. In our algorithm, we have utilized linearly adaptable inertia weight and w_{iter} gives the value of the inertia weight at that given iteration. The iterative process is continued until the optimization process yields a satisfactory result. This is evaluated on the basis of whether the value of f_{fit} falls below the specified maximum allowable value or whether the maximum number of iterations has been reached. A detailed description of the PSO algorithm is available in [40, 41].

N_v	N_b	Z_{bl}	Z_{vl}	Z_{vr}	Z_{br}	P_b	P_v	w_1	w_2	w_3	K

Fig. 7.3. Detailed configuration of each 12-dimensional "particle" employed by PSO. (Reproduced from [44] with permission from the IEEE. ©2007 IEEE.).

In our approach, the objective of the neuro-fuzzy assistance to the EKF based SLAM is to improve the estimation performance as much as possible. This means we should try and minimize the discrepancy between actual covariance and the theoretical covariance of the innovation sequence over the entire set of observation instants, during the movement of the vehicle/robot, as much as possible. Hence the fitness function is formulated on the basis of: *a*) computing the mean-square value of all the diagonal entries of the $\Delta \mathbf{C}_{Innk}$ matrix at any given observation instant, *b*) storing such mean-square values for each observation instant during an on-going iteration and *c*) computing a mean of all such mean-square values for all observation instants at the end of a complete iteration. Mathematically this can be shown as:

$$f_{fit} = \frac{\displaystyle\sum_{n_{obs}=1}^{N_{obs}} \left(\frac{\displaystyle\sum_{j=1}^{J_{C\_nobs}} [\Delta \mathbf{C}_{Innk}(j,j)]^2}{J_{C\_nobs}} \right)}{N_{obs}} \tag{7.36}$$

where N_{obs} denotes the total number of observation instants in a given iteration and J_{C_nobs} denotes the total number of diagonal elements of $\Delta \mathbf{C}_{Innk}$ matrix when the n_{obs}th observation is made.

In the context of adapting a meaningful NFS, the positions of each "particle", at the end of each iteration, are subjected to several constraints. Most of these constraints are implemented to maintain specific shapes chosen for the MFs (usually trapezoidal, which as a special case can become triangular) and also to ensure that there is some overlapping between the stretches of consecutive MFs. Another constraint included is that, for each MF, its control points (starting from left to right) should be chosen in a monotonically nondecreasing fashion. This will ensure that all regions, within the universe of discourse of the input for the NFS, will remain covered by at least one MF. These constraints are implemented as:

$$
\begin{aligned}
&\textbf{IF } (N_b < N_v) \textbf{ THEN } N_b = N_v \textbf{ ENDIF} \\
&\textbf{IF } (Z_{vl} < Z_{bl}) \textbf{ THEN } Z_{vl} = Z_{bl} \textbf{ ENDIF} \\
&\textbf{IF } (Z_{vr} < Z_{vl}) \textbf{ THEN } Z_{vr} = Z_{vl} \textbf{ ENDIF} \\
&\textbf{IF } (Z_{br} < Z_{vr}) \textbf{ THEN } Z_{br} = Z_{vr} \textbf{ ENDIF} \\
&\textbf{IF } (P_v < P_b) \textbf{ THEN } P_v = P_b \textbf{ ENDIF} \\
&\textbf{IF } (N_b < Z_{bl}) \textbf{ THEN } N_b = Z_{bl} \textbf{ ENDIF} \\
&\textbf{IF } (Z_{br} < P_b) \textbf{ THEN } Z_{br} = P_b \textbf{ ENDIF}
\end{aligned}
\tag{7.37}
$$

Another constraint is implemented to signify that the scaling employed in the output layer of the NFS is employed for magnitude scaling only, and hence it cannot be employed for changing polarity. It means that K cannot become negative.

7.4.3 Performance Evaluation

To evaluate the performance of the proposed system, we have considered various environments, which are available in [42]. In fact the packages available in [42] should serve as an excellent platform for learning and analysis of existing Kalman filter and particle filter based SLAM algorithms. Researchers can develop their own algorithms and can compare their performance vis-à-vis these algorithms. Several benchmark environments are available there and we have tested our algorithm in these simulated environments with their associated given vehicle motion model. The environment is usually specified in such a manner where a vehicle/robot is supposed to navigate through some waypoints and in the process should be able to acquire the map of the environment with several configurations of feature/landmark points. In the present scheme, we consider three such environments as specified in [42]. In each case we have the identical scene of ideal robot movement where the robot path is specified by 17 waypoints. However, each environment consists of varied number of landmarks to impose several degrees of complexities and the three environments under consideration consist of 35, 135 and 497 landmarks respectively. The uncertainties in control inputs are specified as: $\sigma_w = 0.3$ m/sec. and $\sigma_s = 3$ deg. An observation step and the associated update step is carried out after eight consecutive prediction steps, identical with the EKF based algorithm in [42]. This follows a popular notion in EKF-based SLAM community, where instead of employing an observation and update step after each prediction step, one computes several consecutive prediction steps, and then takes corrective action by one observation and update step. This helps in reducing the computational burden of the SLAM algorithm. In algo. 7.1, this is indicated by the *Time_for_Observation* flag, which is set TRUE for one iteration, after each 8 successive iterations.

The performance of the proposed system is compared with a conventional EKF-based SLAM system where the **Q** and **R** matrices are kept static throughout the experiment. The proposed algorithm starts with the same **Q** and **R** matrices, but it keeps adapting the **R** matrix according to the proposed scheme. According to the data available from [42], the EKF based algorithm works perfectly when sensor statistics are known as: $\sigma_r = 0.1$ m. and $\sigma_\theta = 1$ deg. First we consider the situation where the sensor statistics are wrongly considered as: $\sigma_r = 2.0$ m. and $\sigma_\theta = 0.1$ deg. In each figure, the firm lines shown in green, depict the actual path traversed by the robot, while the firm lines shown in black, depict the SLAM estimated path traversed based on estimated states of robot poses in each sampling instant or iteration. The stars (*) depict the actual landmark positions, which are stationary in the environment. The crosses (+) depict the positions of these landmarks estimated at the end of the test run. Obviously, the performance of the system will be superior, if the estimated robot path and actual path match as far as possible and the estimated landmark positions and their actual positions coincide as far as possible. Figure 7.4(a) to Fig. 7.4(c) shows the performance of the conventional EKF-based SLAM for three different environment situations. It can be seen that the performance is acceptable when there are small number of landmarks in the environment. However, the performance became really bad when the landmarks became denser and both the estimations of the robot pose at different instants and the map acquired degraded significantly as the EKF estimations are quite distant from the original robot positions and the map situation. Figure 7.5(a) to Fig. 7.5(c) show the situations when the neuro-fuzzy assisted EKF-based SLAM is employed for identical environments. It can be seen that the neuro-fuzzy assistance could improve the situation dramatically and the estimates of the robot states as well as acquisition of the map was quite stable for all three different environments with varied number of landmarks. In all these environments, the robot position estimates follow the actual robot positions closely and the estimation of the stationary landmark positions also closely matches with their actual positions in the environments.

The scheme was further tested for another situation where the sensor statistics are wrongly considered in opposite directions and they are considered as $\sigma_r = 0.01$ m. and $\sigma_\theta = 3.0$ deg. Then the same set of algorithms was employed for identical set of environments. Figure 7.6(a) to Fig. 7.6(c) show the performances of the conventional EKF-based SLAM and Fig. 7.7(a) to Fig. 7.7(c) show the corresponding performances of the neuro-fuzzy assisted EKF-based SLAM algorithms. In these case studies, the EKF-based SLAM shows a different trend in performance. As we can see, the estimation performance is worst for the environment containing small number of landmarks. However, with increase in landmarks, the estimations became more accurate and for the situation with 497 landmarks, the performance of the EKF-based SLAM was quite satisfactory. On the other hand, the neuro-fuzzy assisted EKF showed uniformly stable performance for each environment with quite accurate estimations of robot poses and feature positions for each environment situation. Each result, shown in Fig. 7.5(a) to Fig. 7.5(c) and Fig. 7.7(a) to Fig. 7.7(c), for the neuro-fuzzy assisted EKF based SLAM depicts one sample run conducted. For each of these six specific situations of two case studies, we conducted 10 individual runs. It was found that, for each given situation, results obtained with each of 10 individual runs, were very close to each other. These case studies further prove that the neuro-fuzzy assistance can vastly improve the degrading performance of the traditional EKF algorithm in several situations, when the sensor statistics are wrongly known. In these situations, the performance of the conventional EKF becomes highly unreliable. However, presence of neuro-fuzzy assistance can help the EKF to maintain a stable performance and this performance has been shown robust enough over several environment situations, with several wrong knowledge of sensor statistics.

For the neuro-fuzzy assisted EKF based SLAM, the training of the neuro-fuzzy system, for each case study as described before, was carried out in offline situation on the basis of the data gathered by the robot for a given environment situation. For our experimentation, we implemented the training procedure, for each case study, for the environment containing 135 landmarks. Once the training of the neuro-fuzzy system was completed (on the basis of a given configuration of the landmarks) and the free parameters of the NFS were suitably determined, the trained NFS-based EKF was implemented for robot navigation through the waypoints for several configuration of landmarks as described before (i.e. environments with 35, 135 and 497 landmarks). Table 7.1 details these parameters employed for the PSO algorithm employed for training the NFS. Here, the dimensions of each particle, which are employed to learn the control points of the MFs of the NFS (i.e. $[x_1 \ x_2 \cdots x_8]$), are all initialized with their positions within the range [-1, 1]. This is done in conformation with the normalization procedure that works in conjunction with the NFS. The prospective weights associated with the layer 3 of the NFS (denoted by the dimensions x_9, x_{10} and x_{11} of the PSO algorithm) are all initialized with their positions within the range [-2, 2]. The prospective gain K associated with the layer 4 of the NFS (denoted by the dimension x_{12} of the PSO algorithm) is initialized with its position within the range [0, 2], because it is assumed that K is a non-negative quantity. Each time, the termination criterion for the PSO algorithm was set for a maximum number of iterations (*maxiter*) of 20. For the case study with initial sensor information $\sigma_r = 2.0$ m. and $\sigma_\theta = 0.1$ deg, the learned parameters of the NFS at the completion of the training procedure are:

$[x_1 \ x_2 \cdots x_{12}]$ = [-0.2008 -0.0626 -0.0626 -0.0626 0.0820 0.5961 0.3224 0.4002 -0.0086 1.5801 -0.9729 0.0011]

and for the case study with initial sensor information $\sigma_r = 0.01$ m. and $\sigma_\theta = 3.0$ deg., the learned parameters of the NFS are:

$[x_1 \ x_2 \cdots x_{12}]$ = [-0.4570 0.5242 0.4805 0.9741 0.9741 0.9741 -0.4290 0.2413 -0.0024 -0.8762 1.3561 0.2907].

In each case, it can be seen that these learned parameters satisfied those constraints presented in (7.37).

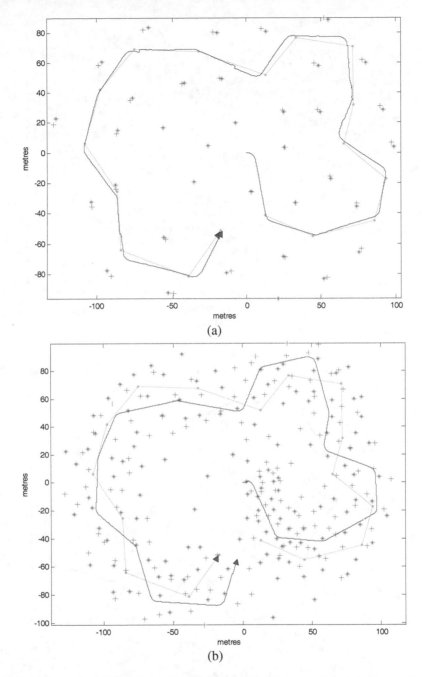

Fig. 7.4. Conventional EKF-based SLAM performance for case study I ($\sigma_r = 2.0$ m. and $\sigma_b = 0.1$ deg.) with (a) 35, (b) 135 and (c) 497 features/landmarks in the environment. (Reproduced from [44] with permission from the IEEE. ©2007 IEEE.).

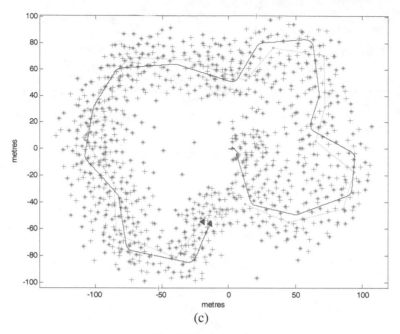

(c)

Fig. 7.4. (*continued*)

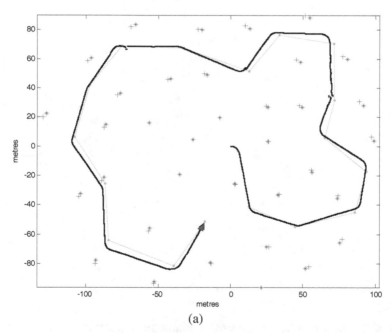

(a)

Fig. 7.5. Neuro-fuzzy assisted EKF-based SLAM performance for case study I ($\sigma_r = 2.0$ m. and $\sigma_b = 0.1$ deg.) with (a) 35, (b) 135 and (c) 497 features/landmarks in the environment. (Reproduced from [44] with permission from the IEEE. ©2007 IEEE.).

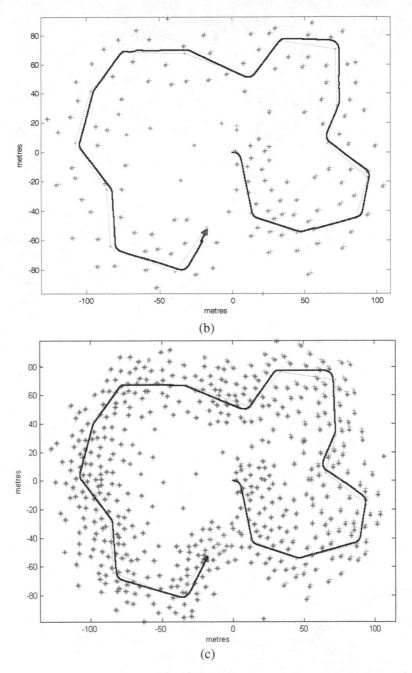

(b)

(c)

Fig. 7.5. (*continued*)

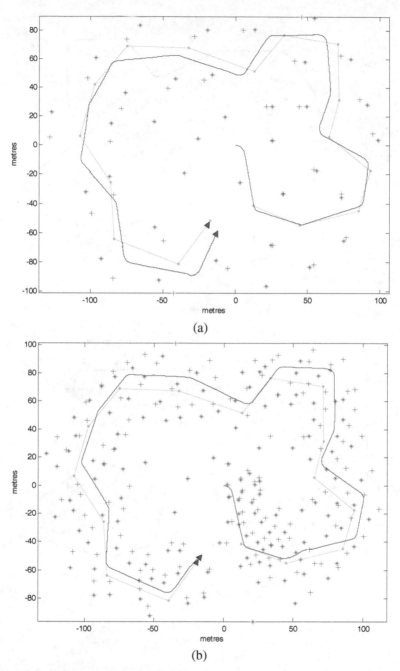

(a)

(b)

Fig. 7.6. Conventional EKF-based SLAM performance for case study II ($\sigma_r = 0.01$ m. and $\sigma_b = 3.0$ deg.) with (a) 35, (b) 135 and (c) 497 features/landmarks in the environment. (Reproduced from [44] with permission from the IEEE. ©2007 IEEE.).

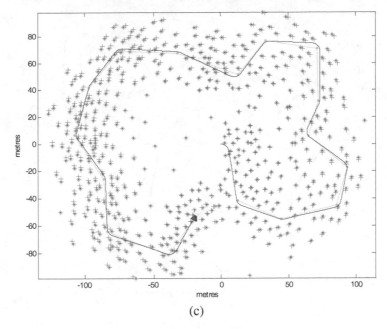

(c)

Fig. 7.6. (*continued*)

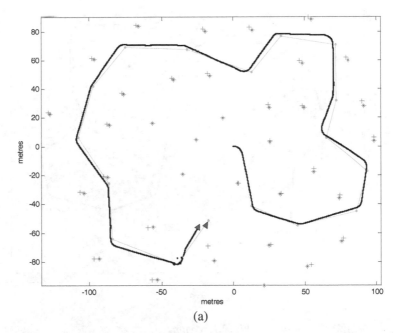

(a)

Fig. 7.7. Neuro-fuzzy assisted EKF-based SLAM performance for case study II ($\sigma_r = 0.01$ m. and $\sigma_b = 3.0$ deg.) with (a) 35, (b) 135 and (c) 497 features/landmarks in the environment. (Reproduced from [44] with permission from the IEEE. ©2007 IEEE.).

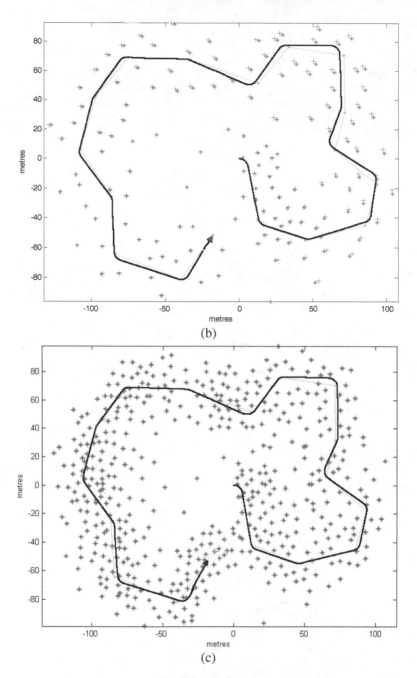

(b)

(c)

Fig. 7.7. (*continued*)

Table 7.1. The PSO parameters employed

Sl. No.	Parameter descriptions	Parameter values for case study (i)	Parameter values for case study (ii)
1	No. of particles (N)	40	40
2	No. of dimensions (D)	12	12
3	Initial inertia weight ($W_{initial}$)	0.9	0.9
4	Slope of inertia weight (ΔW)	2.5e-4	2.5e-4
5	Initialization range for MFs (x_1, x_2, ... x_8)	[-1, 1]	[-1, 1]
6	Initialization range for weight factors (x_9, x_{10}, x_{11})	[-2, 2]	[-2, 2]
7	Initialization range for gain (x_{12})	[0, 2]	[0, 2]
8	Maximum permissible velocity for MFs (v_{1max}, v_{2max}, ... v_{8max})	0.3	0.1
9	Maximum permissible velocity for weight factors (v_{9max}, v_{10max}, v_{11max})	1.0	0.5
10	Maximum permissible velocity for gain (v_{12max})	1.0	0.5

7.5 Training a Fuzzy Supervisor Employing Differential Evolution (DE) Based Optimization

In the previous section we demonstrated how PSO can be utilized to train a fuzzy/neuro-fuzzy supervisor for successful supervision of an EKF based SLAM system. Logically speaking, the idea can be extended to employ other evolutionary algorithms too for similar fuzzy/neuro-fuzzy based supervision purpose. Hence we implemented a similar fuzzy supervisor employing differential evolution (DE), another popular evolutionary algorithm known, for similar types of problems [45]. In DE, like many other population based global optimization methods, several candidate solutions, each containing a possible solution vector for the optimization problem under consideration, are created simultaneously in the multi-dimensional search space and each one of them is individually evaluated in terms of its fitness function, which indicates the degree of suitability of that particular candidate solution to evolve as the best possible solution. This process is continued in an iterative fashion, where new vectors, i.e. possible candidate solutions, are created from the candidate solutions in the previous generation, in quest for generation of better and better solutions, which can be quantitatively evaluated by fitter and fitter fitness function values. Several mathematical strategies can be employed to create new candidate vectors for the current generation, based on the old candidate vectors of the previous generation. At the end of each generation, the candidate solution providing the fittest fitness function value (usually the minimum value) emerges as the best possible solution. This iterative process continues until the fittest fitness function value (usually the minimum value) for the best solution vector in a generation falls below the maximum permitted fitness function value for that optimization process or when the maximum number of generations is reached.

Let us consider that, in the basic variant of DE, utilized for minimizing a cost function $f(\mathbf{x})$ on the basis of D-dimensional $\mathbf{x}$, NP number of such candidate solutions of $(x_1, x_2, \cdots x_D)$ are created in the D-dimensional space and the suitability of each of them is evaluated in each generation G. The initial population is generated in a random fashion and the objective is that the generated vectors should try to cover the entire search space as far as practicable. Each ith vector for the $(G+1)$th generation is created by adding the weighted difference between two population vectors to a third vector, all these three vectors pertaining to the Gth generation. This can be shown by the following formula [46],[47]:

$$v_{i,G+1} = x_{r_1,G} + F(x_{r_2,G} - x_{r_3,G}) \qquad (7.38)$$

where $i = 1, 2, \cdots, NP$. Here $r_1, r_2, r_3 \in [1, NP]$ and they are all mutually different. F is a constant weighting factor and usually $F \in [0, 2]$. This factor influences the amplification of the difference $(x_{r_2,G} - x_{r_3,G})$.

To increase diversity in the newly generated vector, the method of crossover is introduced. This crossover operation generates a new vector $u_{i,G+1}$, from the newly generated perturbed vector $v_{i,G+1}$ and the old vector $x_{i,G}$. In the basic variant of DE, this new vector is generated as [11,12]:

$$u_{i,G+1} = (u_{1i,G+1}, u_{2i,G+1}, \cdots u_{Di,G+1}) \text{ with}$$

$$u_{ji,G+1} = \begin{cases} v_{ji,G+1} & for & j = \langle n+1 \rangle_D \cdots \langle n+L \rangle_D \\ x_{ji,G} & for\ all & other & j \in [1, D] \end{cases} \qquad (7.39)$$

Here, n is a randomly chosen integer, $n \in [1, D]$, and it determines the starting index for the crossover. The length or duration of crossover, in this basic variant of DE, is also an integer drawn from the interval [1,D], and is based on the chosen crossover probability, $CR \in [0, 1]$. These n and L values are chosen afresh for each $u_{i,G+1}$.

Now, if the new vector $u_{i,G+1}$ can yield a smaller value for the fitness function, then this vector becomes the new $x_{i,G+1}$ for the $(G+1)$th generation. Otherwise we keep $x_{i,G+1} = x_{i,G}$.

7.5.1 Performance Evaluation

The performance of DE optimized fuzzy supervisor based solution for the SLAM problems has also been tested by creating an environment in simulation, utilizing the package available in [42], as done in our previous set of case studies. For the new set of case studies, we consider a different environment and two sets of incorrect knowledge of sensor statistics as: (a) $\sigma_r = 0.01$ m. and $\sigma_b = 10.0$ deg. and (b) $\sigma_r = 0.01$ m. and $\sigma_b = 15.0$ deg. For these situations, the performances exhibited by the conventional EKF-based SLAM [42] are shown in Fig. 7.8(a) and Fig. 7.8(b). It can be seen that the estimated robot path deviates a lot from the ideal path and also the estimated positions of many landmarks are quite far away from their actual positions. However, when our DE-optimized fuzzy supervisor based system was employed for each of these two case studies, the fuzzy supervision could improve the performance quite markedly, in each case, as depicted in Fig. 7.9(a) and Fig. 7.9(b). For the fuzzy supervised algorithm, the estimated robot paths deviated much less from the ideal robot paths. In this scheme, the free parameters of the fuzzy supervisor are learnt by implementing differential evolution with $D = 11$ and employing binomial crossover. The variety of the DE algorithm employed is a popular variant, known as the "DE/rand/1" scheme [46], [47]. However, this variant differs slightly from the original "DE/rand/1" scheme, because here the random selection of vectors is performed by shuffling the array containing the population so that a given vector does not get chosen twice in the same term contained in the perturbation expression [48]. It can also be seen that, for each case study, the estimated positions of the landmarks are in closer agreement with their actual positions, than the systems utilizing conventional EKF-based SLAM algorithms.

The results shown in Fig. 7.9 are obtained in the implementation phase, using the fuzzy supervisors trained by the DE algorithm, with the chosen control parameters $NP = 20$, $F = 0.1$, $CR = 0.5$. Like most other stochastic global optimization methods, the performance of the differential evolution strategy too varies with the choice of these free parameters. Hence proper choice or fitting of these parameters is crucial. According to the general guidelines proposed in [46], for many applications, choices of $NP = 10*D$, $F \in [0.5, 1]$ and $CR \in [0, 1]$ but much lower than 1, are considered to be good choices. Among these factors, F is considered to be the most crucial control parameter and NP and CR are considered less crucial ones. Hence, in order to find the best performance of DE, it was considered to carry out simulations for various values of these control parameters and to observe their corresponding performances, for the case study with sensor statistics ($\sigma_r = 0.01$ m. and $\sigma_b = 15.0$ deg.). At first, NP and CR are kept fixed at 20 and 0.5 respectively and varied F for a number of values in the range 0 to 1 and for each case the fuzzy supervisor was trained separately. Although, according to the general guideline NP should have been chosen as $10*11=110$, this would have increased the computational burden of the training procedure enormously. Hence, with the objective of keeping the computational burden reasonably low, the optimization procedure was attempted with an NP value of 20. Here when F was varied, it was found that better and better performance of the overall system could be achieved in the implementation phase if we use smaller values of F. It was found that the best performance was achieved with $F = 0.1$ and with lower values of F the performance degraded a little while with higher values of F the degradation was significant. Figure 7.10(a) to Fig. 7.10(c) show the corresponding performances of the system in the implementation phase with the trained fuzzy supervision for $F = 0.05$, $F = 0.1$ and $F = 0.5$. Figure 7.11 shows the RMS errors in estimating $\hat{x}$, in the implementation phase, at each sampling instant with an incremental movement of the robot, for this series of case studies with five representative values of F. It can be easily concluded that the training process conducted with $F = 0.1$ produced the best result for these experimentations.

With this value of F, then one can proceed to determine the most suitable values of NP and CR. Keeping $F = 0.1$ and $CR = 0.5$, we varied NP for a series of values. The objective was to obtain a reasonable performance with as small a value of NP as practicable, so that the computational burden is kept minimum. Figure 7.12 shows the RMS errors in estimating $\hat{\mathbf{x}}$, in the implementation phase, at each sampling instant with an incremental movement of the robot, for this series of case studies with three representative values of NP = 15, 20 and 25. It was found that the best performance is obtained with NP = 20 and the performance degrades if we either increase or decrease the value of NP. Hence a value of NP = 20 was chosen for the training procedure. Next keeping F = 0.1 and NP = 20, CR was varied for a series of values. It was found that the variation of CR was not that critical in varying the training performance of the scheme. Figure 7.13 shows the similar plotting of RMS errors in estimating $\hat{\mathbf{x}}$, for this series of case studies with three representative values of CR = 0.4, 0.5 and 0.6. It was found that the best performance was obtained with CR = 0.5 although performances for other values of CR were quite similar in nature. Hence it could be concluded that the best set of control parameters of the DE for the training procedure of the fuzzy supervisor is obtained as NP = 20, F = 0.1 and CR = 0.5. Hence, using these parameters the fuzzy supervisor was trained for each case study of sensor statistics i.e. (a) with (σ_r = 0.01 m. and σ_b = 10.0 deg.) and (b) with (σ_r = 0.01 m. and σ_b = 15.0 deg.). Figure 7.9(a) and Fig. 7.9(b) showed the performances of those case studies, in the implementation phase.

In the next phase, we present a performance comparison between the fuzzy supervisor tuned by DE and the fuzzy supervisor tuned by PSO. The performance comparison is demonstrated for the sample case study with sensor statistics (σ_r = 0.01 m. and σ_b = 15.0 deg.). The popular version of PSO, employed using linearly decreasing inertia weight, as described in (7.35), is used for this purpose. To make as uniform comparison between the DE based and the PSO based tuning algorithms for our problem as practicable, the following factors are taken into consideration: (i) identical number of candidate solutions or particles for each algorithm (i.e. 20), (ii) identical value of maximum number of iterations or generations for which the optimization algorithm is run each time (taken as 10 in this work) and (iii) identical range of initialization of each corresponding dimension of the initial population for each optimization algorithm. The PSO with inertia weight variation is normally known to perform well for benchmark optimization functions with initial inertia weight, $W_{initial}$, of 0.9 and slope of inertial weight of 2.5e-4. For our case study, we implemented PSO with $W_{initial}$ = 0.9 and employed a series of both slow decrease and aggressive decrease in inertia weight. Figure 7.14 shows the corresponding performance of the PSO algorithm in terms of the RMS errors in estimating $\hat{\mathbf{x}}$, in the implementation phase, at each sampling instant with an incremental movement of the robot, for this series of case studies when the PSO-based training procedure was conducted with slope of inertia weight having values 2.0e-4, 2.5e-4, 5.0e-4, 4e-2 and 5e-2. It was found that the best performance was indeed obtained with the universally known superior value of 2.5e-4. Figure 7.15 shows a similar comparison of estimation performance for the best PSO-tuned and best DE-tuned fuzzy supervisors for the

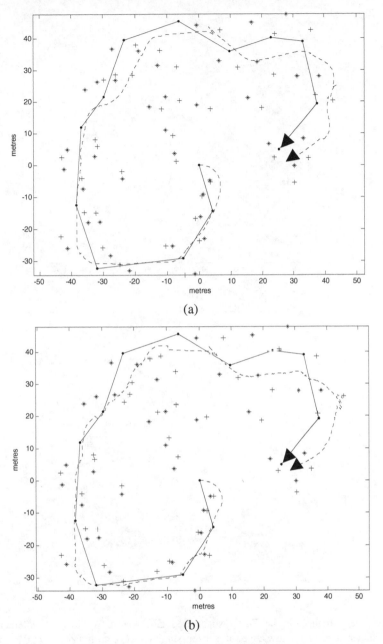

Fig. 7.8. Performance of the conventional EKF-based SLAM under incorrect knowledge of sensor statistics: (a) with ($\sigma_r = 0.01$ m. and $\sigma_b = 10.0$ deg.) and (b) with ($\sigma_r = 0.01$ m. and $\sigma_b = 15.0$ deg.)

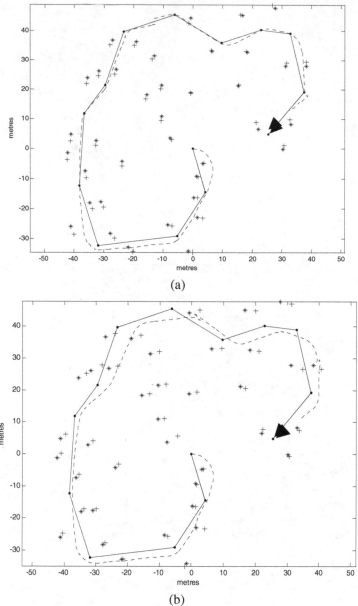

(a)

(b)

Fig. 7.9. Performance of the Fuzzy supervised EKF-based SLAM, in implementation phase, under incorrect knowledge of sensor statistics: (a) with ($\sigma_r = 0.01$ m. and $\sigma_b = 10.0$ deg.) and (b) with ($\sigma_r = 0.01$ m. and $\sigma_b = 15.0$ deg.)

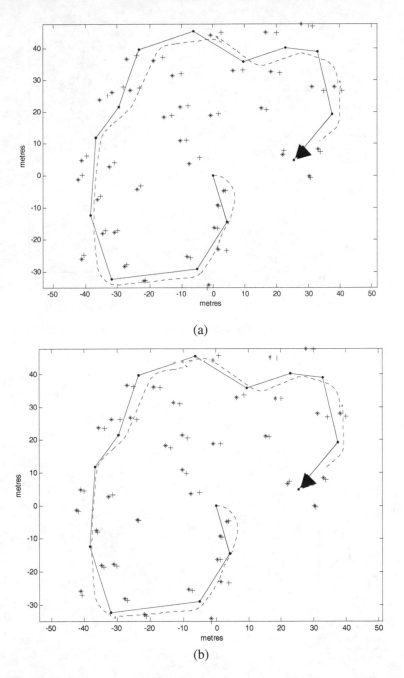

(a)

(b)

Fig. 7.10. The implementation performance of the fuzzy supervised EKF-based SLAM, when the DE-based training was carried out with $NP = 20$, $CR = 0.5$, and (a) $F = 0.05$, (b) $F = 0.1$, and (c) $F = 0.5$

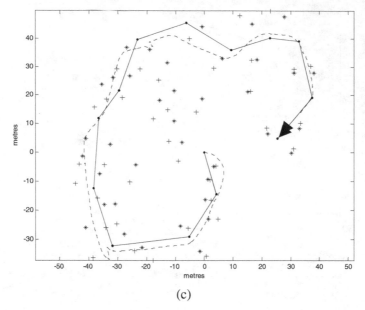

(c)

Fig. 7.10. (*continued*)

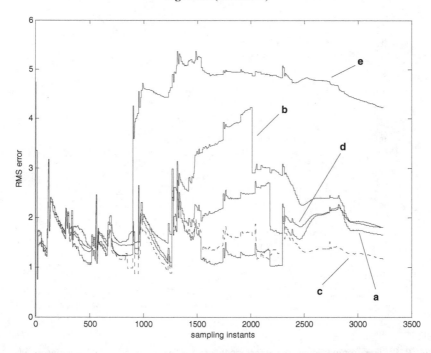

Fig. 7.11. The estimation performance of the fuzzy supervised EKF-based SLAM, in the implementation phase, when the DE-based training was carried out with $NP = 20$, $CR = 0.5$, and (a) $F = 0.05$, (b) $F = 0.08$, (c) $F = 0.1$, (d) $F = 0.15$, and (c) $F = 0.5$

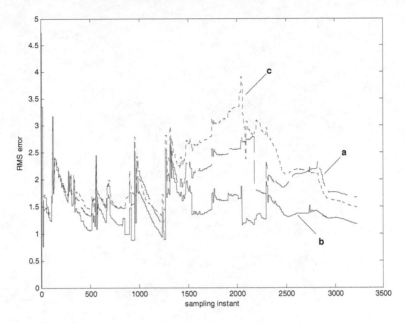

Fig. 7.12. The estimation performance of the fuzzy supervised EKF-based SLAM, in the implementation phase, when the DE-based training was carried out with $F = 0.1$, $CR = 0.5$, and (a) $NP = 15$, (b) $NP = 20$, and (c) $NP = 25$

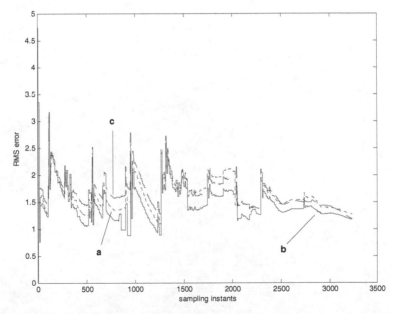

Fig. 7.13. The estimation performance of the fuzzy supervised EKF-based SLAM, in the implementation phase, when the DE-based training was carried out with $F = 0.1$, $NP = 20$, and (a) $CR = 0.4$, (b) $CR = 0.5$, and (c) $CR = 0.6$

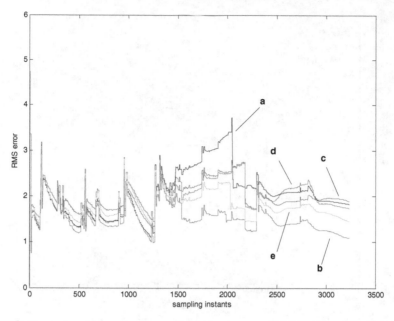

Fig. 7.14. The estimation performance of the fuzzy supervised EKF-based SLAM, in the implementation phase, when the PSO-based training was carried out with the slope of inertia weight chosen as (a) 2.0e-4, (b) 2.5e-4, (c) 5.0e-4, (d) 4e-2, and (e) 5e-2

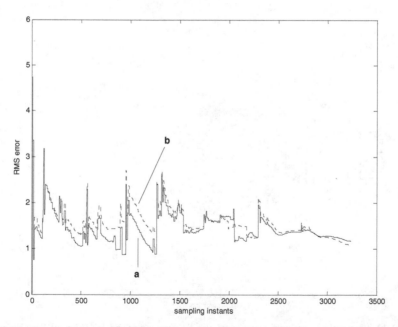

Fig. 7.15. Comparison of the estimation performance of the fuzzy supervised EKF-based SLAM, in the implementation phase, when the fuzzy supervisor is trained by (a) DE algorithm and (b) PSO algorithm

adaptive EKF based SLAM algorithm, for the case study under consideration. It can be seen that the performance of the DE tuned algorithm gave less RMS errors in estimation, at most of the sampling instants. This procedure helps us demonstrating the usefulness of employing a DE-tuned fuzzy supervision for EKF based SLAM problems. However we would like to generally remark that this performance may vary depending on the environment chosen and the sensor statistics considered.

7.6 Summary

The present chapter discussed the importance of SLAM in the context of mobile robot navigation and, at first, described the extended Kalman filter based SLAM algorithms in detail. Next we considered the degradation in system performance when *a priori* knowledge of the sensor statistics is incorrect and showed how fuzzy/neuro-fuzzy assistance or supervision can significantly improve the performance of the algorithm. Usually, EKF is known as a good choice for SLAM algorithms when the associated statistical models are well known. However, the performance can become significantly unpredictable and degrading when the knowledge of such statistics is inappropriate. The fuzzy/neuro-fuzzy supervisor based system proposes to start the system with the wrongly known statistics and then adapt the **R** matrix, online, on the basis of a fuzzy/neuro-fuzzy system that attempts to minimize the mismatch between the theoretical and the actual values of the innovation sequence. The free parameters of the neuro-fuzzy system are automatically learned employing an evolutionary optimization based training procedure. The chapter showed how two popular contemporary evolutionary optimization techniques, namely, PSO and DE, can be utilized successfully for this purpose. The performance evaluation is carried out for several benchmark environment situations with several wrong knowledge of sensor statistics. While the conventional EKF based SLAM showed unreliable performance with significant degradation in many situations, the fuzzy/neuro-fuzzy assistance could improve this EKF's performance significantly and could provide robust, accurate performance in each sample situation in each case study.

Acknowledgement. This work was partially supported by JSPS Postdoctoral Fellowship for Foreign Researchers in Japan. This work was also partially supported by All India Council for Technical Education under RPS scheme (Grant No. 8023/BOR/RPS-89/2006-07).

References

[1]   Dissanayake, M.W.M.G., Newman, P., Clark, S., Durrant-Whyte, H.F.: A solution to the simultaneous localization and map building (SLAM) problem. IEEE Tran. Robotics and Automation 17(3), 229–241 (2001)

[2]   Montemerlo, M., Thrun, S., Koller, D., Wegbreit, B.: FastSLAM 2.0: An improved particle filtering algorithm for simultaneous localization and mapping that provably converges. In: Proc. 18th International Joint Conference on Artificial Intelligence (IJCAI), Acapulco, Mexico (2003)

[3]   Grisetti, G., Stachniss, C., Burgard, W.: Improving grid-based SLAM with Rao-Blackwellized particle filters by adaptive proposals and selective resampling. In: Proc. of the IEEE International Conference on Robotics and Automation (ICRA), Barcelona, Spain, pp. 2443–2448 (2005)

[4]   Smith, R., Cheeseman, P.: On the representation and estimation of spatial uncertainty. International Journal of Robotics Research 5(4) (1986)

[5]   Moutarlier, P., Chatila, R.: Stochastic multisensory data fusion for mobile robot location and environment modeling. In: 5th Int. Symposium on Robotics Research, Tokyo (1989)

[6]   Davison, A.J.: Mobile Robot Navigation Using Active Vision. PhD Thesis, Univ. of Oxford (1998)

[7]   Bailey, T.: Mobile Robot Localization and Mapping in Extensive Outdoor Environments. PhD Thesis, Univ. of Sydney (2002)

[8]   Davison, A.J., Murray, D.W.: Simultaneous localization and map-building using active vision. IEEE Tran. Pattern Analysis and Machine Intelligence 24(7), 865–880 (2002)

[9]   Guivant, J., Nebot, E.: Optimization of the simultaneous localization and map-building algorithm and real-time implementation. IEEE Tran. Robotics and Automation 17(3), 242–257 (2001)

[10] Guivant, J., Nebot, E.: Solving computational and memory requirements of feature-based simultaneous localization and mapping algorithms. IEEE Tran. Robotics and Automation 19(4), 749–755 (2003)

[11] Williams, S.B., Newman, P., Dissanayake, G., Durrant-Whyte, H.: Autonomous underwater simultaneous localization and map building. In: Proc. IEEE International Conference on Robotics and Automation, San Francisco, CA, vol. 2, pp. 1792–1798 (2000)

[12] Chong, K.S., Kleeman, L.: Feature-based mapping in real, large scale environments using an ultrasonic array. International Journal of Robotic Research 18(2), 3–19 (1999)

[13] Bosse, M., Leonard, J., Teller, S.: Large-scale CML using a network of multiple local maps. In: Leonard, J., Tardós, J.D., Thrun, S., Choset, H. (eds.) Workshop Notes of the ICRA Workshopon Concurrent Mapping and Localization for Autonomous Mobile Robots (W4), Washington, DC. ICRA Conference (2002)

[14] Thrun, S., Fox, D., Burgard, W.: A probabilistic approach to concurrent mapping and localization for mobile robots. Machine Learning 31, 29–53 (1998); also appeared in Autonomous Robots 5, 253–271 (joint issue)

[15] Williams, S., Dissanayake, G., Durrant-Whyte, H.F.: Towards terrain-aided navigation for underwater robotics. Advanced Robotics 15(5) (2001)

[16] Thrun, S., Hähnel, D., Ferguson, D., Montemerlo, M., Triebel, R., Burgard, W., Baker, C., Omohundro, Z., Thayer, S., Whittaker, W.: A system for volumetric robotic mapping of abandoned mines. In: Proceedings of the IEEE International Conference on Robotics and Automation, ICRA (2003)

[17] Castellanos, J.A., Montiel, J.M.M., Neira, J., Tardós, J.D.: The SPmap: A probabilistic framework for simultaneous localization and map building. IEEE Transactions on Robotics and Automation 15(5), 948–953 (1999)

[18] Paskin, M.A.: Thin junction tree filters for simultaneous localization and mapping. In: Proceedings of the Sixteenth International Joint Conference on Artificial Intelligence (IJCAI), Acapulco, Mexico (2003)

[19]   Thrun, S., Koller, D., Ghahramani, Z., Durrant-Whyte, H., Ng, A.Y.: Simultaneous mapping and localization with sparse extended information filters. In: Boissonnat, J.-D., Burdick, J., Goldberg, K., Hutchinson, S. (eds.) Proceedings of the Fifth International Workshop on Algorithmic Foundations of Robotics, Nice, France (2002)

[20]   Neira, J., Tardós, J.D.: Data association in stochastic mapping using the joint compatibility test. IEEE Transactions on Robotics and Automation 17(6), 890–897 (2001)

[21]   Shatkay, H., Kaelbling, L.: Learning topological maps with weak local odometric information. In: Proceedings of IJCAI 1997. IJCAI, Inc. (1997)

[22]   Araneda, A.: Statistical inference in mapping and localization for a mobile robot. In: Bernardo, J.M., Bayarri, M.J., Berger, J.O., Dawid, A.P., Heckerman, D., Smith, A.F.M., West, M. (eds.) Bayesian Statistics 7. Oxford University Press, Oxford (2003)

[23]   Montemerlo, M., Thrun, S.: Simultaneous localization and mapping with unknown data association using Fast SLAM. In: Proc. IEEE International Conference on Robotics and Automation (ICRA), Taipei, Taiwan (2003)

[24]   Hu, W., Downs, T., Wyeth, G., Milford, M., Prasser, D.: A modified particle filter for simultaneous robot localization and Landmark tracking in an indoor environment. In: Proc. Australian Conference on Robotics and Automation (ACRA), Canberra, Australia (2004)

[25]   Frese, U., Larsson, P., Duckett, T.: A multilevel relaxation algorithm for simultaneous localization and mapping. IEEE Tran. Robotics 21(2), 196–207 (2005)

[26]   Montemerlo, M., Thrun, S., Koller, D., Wegbreit, B.: FastSLAM: A factored solution to the simultaneous localization and mapping problem. In: Proceedings of the AAAI National Conference on Artificial Intelligence, Edmonton, Canada, AAAI (2002)

[27]   Lu, F., Milios, E.: Globally consistent range scan alignment for environment mapping. Autonomous Robots 4, 333–349 (1997)

[28]   Mehra, R.K.: On the identification of variances and adaptive Kalman filtering. IEEE Tran. Automatic Control AC-15(2), 175–184 (1970)

[29]   Fitzgerald, R.J.: Divergence of the Kalman filter. IEEE Tran. Automatic Control AC-16(6), 736–747 (1971)

[30]   Sinha, N.K., Tom, A.: Adaptive state estimation for systems with unknown noise covariances. International Journal of Systems Science 8(4), 377–384 (1977)

[31]   Bellanger, P.R.: Estimation of noise covariance matrices for a linear time-varying stochastic process. Automatica 10, 267–275 (1974)

[32]   Dee, D.P., Cohn, S.E., Dalcher, A., Ghil, M.: An efficient algorithm for estimating noise covariances in distributed systems. IEEE Tran. Automatic Control AC-30(11), 1057–1065 (1985)

[33]   Reynolds, R.G.: Robust estimation of covariance matrices. IEEE Tran. Automatic Control 32(9), 1047–1051 (1990)

[34]   Morikawa, H., Fujisaki, H.: System identification of the speech production process based on a state-space representation. IEEE Trans. Acoust., Speech, Signal Processing ASSP-32, 252–262 (1984)

[35]   Noriega, G., Pasupathy, S.: Adaptive estimation of noise covariance matrices in real-time preprocessing of geophysical data. IEEE Trans. Geoscience and Remote Sensing 35(5), 1146–1159 (1997)

[36] Kobayashi, K., Cheok, K.C., Watanabe, K., Munekata, F.: Accurate differential global positioning system via fuzzy logic Kalman filter sensor fusion technique. IEEE Tran. Industrial Electronics 45(3), 510–518 (1998)

[37] Loebis, D., Sutton, R., Chudley, J., Naeem, W.: Adaptive tuning of a Kalman filter via fuzzy logic for an intelligent AUV navigation system. Control Engineering Practice 12, 1531–1539 (2004)

[38] Wu, Z.Q., Harris, C.J.: An adaptive neurofuzzy Kalman filter. In: Proc. 5th International Conference on Fuzzy Sets and Systems FUZZ-IEEE 1996, vol. 2, pp. 1344–1350 (September 1996)

[39] Sasiadek, J.Z., Wang, Q., Zeremba, M.B.: Fuzzy adaptive Kalman filtering for INS/GPS data fusion. In: Proc. 15th International Symposium on Intelligent Control (ISIC 2000), Rio, Patras, Greece (July 2000)

[40] Clerc, M., Kennedy, J.: The particle swarm-explosion, stability and convergence in a multidimensional complex space. IEEE Tran. Evolutionary Computation 6(1), 58–73 (2002)

[41] Shi, Y., Eberhart, R.C.: Empirical study of particle swarm optimization. In: Proceedings of the 1999 Congr. Evolutionary Computation, pp. 1945–1950. IEEE Service Center, Piscataway (1999)

[42] http://www.acfr.usyd.edu.au/homepages/academic/tbailey/software/software.html

[43] Brown, R.G., Hwang, P.Y.C.: Introduction to Random Signals and Applied Kalman Filtering, 3rd edn. John Wiley and Sons, USA (1997)

[44] Chatterjee, A., Matsuno, F.: A neuro-fuzzy assisted extended Kalman filter-based approach for Simultaneous Localization and Mapping (SLAM) problems. IEEE Transactions on Fuzzy Systems 15(5), 984–997 (2007)

[45] Chatterjee, A.: Differential evolution tuned fuzzy supervisor adapted extended kalman filtering for SLAM problems in mobile robots. Robotica 27(3), 411–423 (2009)

[46] Storn, R.: On the usage of differential evolution for function optimization (1996)

[47] Storn, R., Price, K.: Minimizing the real functions of the ICEC 1996 contest by differential evolution (1996)

[48] http://www.icsi.berkeley.edu/~storn/code.html (last accessed June 24, 2008)

Chapter 8
Vision Based SLAM in Mobile Robots[*]

Abstract. This chapter is an extension of the previous chapter and it discusses how the previously discussed concept of SLAM for mobile robots can be actually implemented in real-life in an indoor environment. The system developed employs a two camera based vision system which successfully performs image feature identification and tracking.

8.1 Introduction

As mentioned in the previous chapter, the extended Kalman filter (EKF) based approach has been widely regarded as probably the most suitable approach for solving the simultaneous localization and mapping (SLAM) problem for mobile robots [1-7]. The basic strength of EKF in solving the SLAM problem lies in its iterative approach of determining the estimation and hence building of an augmented map of its surrounding environment through which the robot is directed to navigate through some waypoints. Here we assume that both the initial localization of the robot pose and the map to be built is unknown to us and we gradually build the map by considering it as an augmentation of estimated states, which are nothing but a collection of the positions of the features or landmarks in the environment, along with the robot's pose. The estimations of these states are integrally associated with some uncertainties in these estimates and they are stored in the form of error covariance matrices. This EKF based SLAM algorithm has been discussed in detail in the previous chapter. In this chapter we shall now discuss how SLAM can be implemented in mobile robots employing vision based sensing.

It is also well regarded that the real implementation of SLAM algorithm for practical environments to build meaningful maps is a difficult task. The accuracy of such a system largely depends on the sensors employed. As we already know, the wheel sensors suffer from wheel-slippage, sonar sensors are low resolution, not highly accurate systems, which also suffer from environmental disturbances,

---

[*] This chapter is adopted from Expert Systems with Applications, vol. 38, issue 7, July 2011, Avishek Chatterjee, Olive Ray, Amitava Chatterjee, and Anjan Rakshit, "Development of a Real-Life EKF based SLAM System for Mobile Robots employing Vision Sensing," pp. 8266-8274, © 2011, with permission from Elsevier.

A. Chatterjee et al.: Vision Based Autonomous Robot Navigation, SCI 455, pp. 207–222.
springerlink.com   © Springer-Verlag Berlin Heidelberg 2013

infra red sensors can only be employed for short distances, laser range finders are expensive and slow in operation due to low update rate and the performance of GPS can suffer due to occlusion of line-of-sight to satellites and their accuracy and update rate may be slow. Hence, solid-state cameras and computers have emerged in recent times as an attractive, feasible, real-time solution for building such robot localization systems [3, 5]. They can also provide comparatively cheaper solution and they can provide great flexibility in interpreting the environment through which a robotic platform is needed to navigate. However, till date, not many works have been reported utilizing vision sensing based SLAM algorithms. The primary reason for that can be that the development of such systems and to make them meaningfully accurate in real-life is essentially a difficult task.

The present chapter will give a detail description of a successful real-life implementation of SLAM algorithm for map development in an indoor environment [15], utilizing a popular differential drive mobile robot, called KOALA robot, which has also been described in previous chapters. An important highlighting feature of the developed scheme is that this stand-alone system utilizes a computer vision based sensing system for building the map. A two-camera based vision system is utilized to perform feature identification, in frames grabbed, and track these features in subsequent frames. Such a system is essential for scene identification and obstacle recognition for a vision-based system that helps in developing suitable navigational algorithms, performing obstacle avoidance and/or developing a map of the environment where the robot is intended to carry out the navigation job. The feature tracking approach is based on minimization of the sum of squared intensity differences between the past and the current window, which determines whether a current window is a warped version of the past window. The system is also equipped with the 3D distance calculation module of the landmarks from the robot frame, which enables to determine the map of the location, storing current localization of the robot along with the co-ordinates of the landmarks in the map. The system has been implemented in real-life in our laboratory for waypoint-directed map development and the system could demonstrate high accuracy in map development in such indoor environments.

8.2 The Dynamic State Model for the Differential Drive Koala Robot

The details of the EKF based SLAM algorithm were already presented in section 7.2. Now, to adapt this theory in the context of the KOALA robot, at first, the dynamic model is developed for the differential-drive based KOALA robot in this section. This can also be logically extended to other similar types of mobile robots too. Here, there are two independent variables governing motion of the vehicle i.e. rotation of the left wheel of the motor and rotation of the right wheel of the motor. However, we consider two derived variables as primary variables and these are (i) linear translation of the geometric center of the robot and (ii) its rotation around the vertical axis through the geometric center. The rationale behind this domain changeover is because of the reason that an error is introduced if we choose 'rotation' as a variable, because of the severe deformation of tier during rotation. Such a problem will not arise in case of linear, translational motion, where the sources of errors or uncertainties are different e.g. incorrect calibration of wheel encoder, small slippage in wheel rotation etc. Here we assume that the robot will never be subjected to simultaneous commands of rotational motion and translational motion.

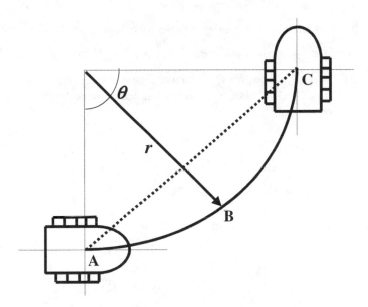

Fig. 8.1. Schematic of the KOALA robot movement

While developing the model, we should keep in mind that the robot always moves along a circular arc. The curvature is zero for linear, translational motion and the radius of curvature is zero for pure rotation. Figure 8.1 shows the schematic of a robot movement. Here

$$\widehat{A}\,\widehat{B}\,\widehat{C} = s \tag{8.1}$$

$$\theta = \frac{s}{r} = K_1 \text{ (Rotation of right wheel – Rotation of left wheel)} \tag{8.2}$$

$$s = (K_2/2) \text{ (Rotation of right wheel + Rotation of left wheel)} \tag{8.3}$$

(8.2) and (8.3) enable us to obtain s and θ directly from the readings of the wheel encoders. Hence we obtain, $r = \dfrac{s}{\theta}$ and $\overline{AC} = 2r\sin\dfrac{\theta}{2}$. Then AC can be decomposed into its x- and y-components, when the initial pose ϕ of the robot is known. Therefore we have:

$$\left.\begin{array}{l} dx = 2r\sin\dfrac{\theta}{2}\cos\phi \\[2mm] dy = 2r\sin\dfrac{\theta}{2}\sin\phi \\[2mm] d\phi = \theta \end{array}\right\} \tag{8.4}$$

The development of such a model gives rise to a logical problem under those situations when $\theta \to 0°$, because then $r = \dfrac{s}{\theta} \to \infty$. Hence, for $\theta < 5°$, it is assumed that $\overline{AC} = s$. Now, for D amount of linear displacement and θ amount of rotation of the KOALA robot, the dynamic model can be finalized using the following formulae:

$$\left.\begin{array}{l} \Delta x = D\cos\phi \\ \Delta y = D\sin\phi \\ \Delta\phi = \theta \end{array}\right\} \Rightarrow \left.\begin{array}{l} x_{k+1} = x_k + D\cos\phi \\ y_{k+1} = y_k + D\sin\phi \\ \phi_{k+1} = \phi_k + \theta \end{array}\right\} \tag{8.5}$$

Hence the Jacobians and the covariance matrix will be calculated as:

$$\nabla f_u = \begin{bmatrix} \dfrac{\partial\Delta x}{\partial D} & \dfrac{\partial\Delta x}{\partial \theta} \\[2mm] \dfrac{\partial\Delta y}{\partial D} & \dfrac{\partial\Delta y}{\partial \theta} \\[2mm] \dfrac{\partial\Delta\phi}{\partial D} & \dfrac{\partial\Delta\phi}{\partial \theta} \end{bmatrix} = \begin{bmatrix} \cos\phi & 0 \\ \sin\phi & 0 \\ 0 & 1 \end{bmatrix} \tag{8.6}$$

$$\nabla f_{x_v} = \begin{bmatrix} 1 & 0 & -D\sin\phi \\ 0 & 1 & D\cos\phi \\ 0 & 0 & 1 \end{bmatrix} \tag{8.7}$$

and $\mathbf{Q} = \begin{bmatrix} \sigma D^2 & 0 \\ 0 & \sigma\theta^2 \end{bmatrix}$ where $\sigma D = D \times$ standard deviation for per unit displacement and $\sigma\theta = \theta \times$ standard deviation for per unit rotation.

8.3 Vision Sensing Based Image Feature Identification, Feature Tracking and 3d Distance Calculation for Each Feature

In our SLAM algorithm, the "observe" step is carried out using vision sensing. The basic version of the KOALA robot is originally procured with some built-in sensors, e.g. incremental wheel encoders and infrared (IR) sensors, and it has been later integrated with several accessories e.g. ultrasonic sensors, wireless radio modem, sensor scanning-tilt-pan system, vision system, servo motors for controlling four degrees of freedom, computing platform etc. All the integrations have been carried out in-house in our laboratory. Figure 8.2 shows the KOALA robot in its integrated form, used specifically for the purpose of performing vision based SLAM.

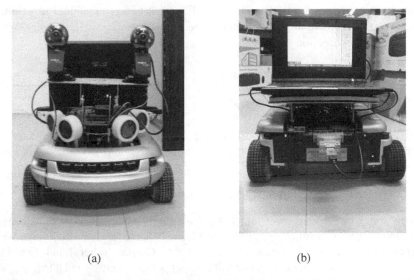

(a)   (b)

Fig. 8.2. KOALA mobile robot, original procured with some built-in sensors, and later integrated in our laboratory with several accessories

The vision-based sensing employs two webcams, as shown in Fig. 8.2(a), for real-life implementation, where the main objective is to implement a two camera based vision system for image feature selection, tracking of the selected features and the calculation of 3D distance of the selected features [16]. This feature identification is based on selection of suitable, candidate image patches or windows in captured frames from running videos acquired from each camera, that have high potential of tracking in subsequent frames. In real life, image patches having high edge information content are better candidates for tracking and hence such patches (considered as static in our system) are considered the best candidate landmarks for developing subsequent maps. The computation of correspondences between features in different views (for our system, the left snap and the right snap i.e. the frames grabbed from the left camera and the right camera) is a necessary precondition to obtain depth information. The system first performs a feature identification algorithm in the frame grabbed from the left camera to identify some suitable rectangular patches or windows that are most suitable as trackable features (patches with sufficient texture) and then it attempts to track them in the frame grabbed from the right camera. The inspiration for developing such a image tracking system is obtained from the Kanade-Lucas-Tomasi (KLT) Tracker [10, 13]. It is always preferable to track a window or patch of image pixels instead of a single pixel because it is almost impossible to track a single pixel, unless it has a very distinctive brightness with respect to all its neighbors. At the same time the result can be confusing, because the intensity value of the pixel can also change due to noise. Hence N number of feature windows is selected, based on the intensity profile, by maintaining a minimum distance between the features in an image frame. For an image f(x, y), a two dimensional function, its gradient is a vector and the gradient of each window G is calculated along x-direction and y-direction as:

$$G = \begin{bmatrix} g_{xx} & g_{xy} \\ g_{xy} & g_{yy} \end{bmatrix} = \begin{bmatrix} g_x^2 & g_x g_y \\ g_x g_y & g_y^2 \end{bmatrix} \tag{8.8}$$

The suitability of the choice of a window as a feature window is evaluated by computing the eigenvalues λ_1 and λ_2 of its G matrix and a feature window is accepted if

$$\min (\lambda_1, \lambda_2) > \lambda \tag{8.9}$$

where λ is a predefined threshold [14]. Two small eigenvalues mean a roughly constant intensity profile within the window. A large and a small eigenvalue correspond to a unidirectional texture pattern. On the other hand two large eigenvalues represent the corners or salt and pepper type texture [11][16].

   Once the features are selected, the next job is to follow or track these features from one frame to another frame in an image sequence [11-13]. Similar to [11], we compute the displacement $\mathbf{dp} = [dxp\ dyp]^T$ of the center of a feature window that minimizes the sum of the squared difference in image intensities between the windows of the two image frames under consideration. In case of the small inter-frame motion, the motion of the features within two image frames can be approximated sufficiently accurately by a pure translation model. However, for bigger inter-frame motions, an affine model, comprising linear warping combined with pure translation, is known to provide better models. Here, the quality of the feature monitored during tracking is better with a dissimilarity measure that includes a deformation matrix that represents the linear warping based affine motion model as well as translations of feature within the frame. The point motion in the image can be described by

$$J(\mathbf{Axp} + \mathbf{dp}) = I(\mathbf{xp}) \tag{8.10}$$

where, J is the current image, I is the original image, $\mathbf{A} = \mathbf{1}+\mathbf{D}$ ($\mathbf{1}$ is a 2x2 identity matrix and $\mathbf{D}$ is the deformation matrix) and $\mathbf{dp}$ is the translation vector. Hence the dissimilarity can be computed utilizing $w(\mathbf{xp})$, a weighting function (popularly chosen as unity or a Gaussian function to emphasize the central portion of the window) as [11]

$$\varepsilon = \iint\limits_{W} [J(\mathbf{Axp} + \mathbf{dp}) - I(\mathbf{xp})]^2\, w(\mathbf{xp})\mathrm{dxp} \tag{8.11}$$

The Newton-Raphson minimization between image intensities of two windows is employed to search for the new position of the center point of a feature window in a new frame in an iterative manner. The following system is needed to be solved to obtain $\mathbf{dp}$:

$$\mathbf{Gdp=e} \tag{8.12}$$

where $\mathbf{G} = \int \left(\mathbf{gg}^{T} w\right)da$; $\mathbf{G}$ = second order weighted coefficient matrix (2×2), $\mathbf{e}$ = weighted intensity error vector (2×1) ($\mathbf{e} = (\int\limits_{W}(I - J)\, \mathbf{g}wda)$, $\mathbf{dp}$ = displacement vector (2×1) ($\mathbf{dp} = [dxp\ \ dyp]^{T}$), and $\mathbf{g}$ = Gradient vector (2×1) ($g = \left[\dfrac{\partial I}{\partial x}\ \ \dfrac{\partial I}{\partial y}\right]^{T}$).

   This iterative algorithm solves (8.12) by solving, in each iteration, for $\begin{bmatrix} g_{xx} & g_{xy} \\ g_{xy} & g_{yy} \end{bmatrix}\begin{bmatrix} dxp \\ dyp \end{bmatrix} = \begin{bmatrix} e_x \\ e_y \end{bmatrix}$ and calculating the new window center in the image, where we are trying to perform the tracking, in that iteration, as $x_{\text{p\_tracked}} = x_{\text{p\_tracked}} + dxp;\ y_{\text{p\_tracked}} = y_{\text{p\_tracked}} + dyp$.

   The 3D distance of the tracked landmarks can be obtained on the basis of data available about the geometry of the camera and the head used [3], [9], [14]. To get depth information in stereo vision, it is required that two lines of sight for the two cameras intersect at a scene point P and from this information the three-dimensional coordinates of the observed scene point in the world co-ordinate system (WCS) can be obtained. Our distance calculation module is based on the pin-hole camera model used in Andrew J. Davison's work [3]. It makes use of the well known camera calibration matrix and perspective projection equation and utilizes the "Midpoint of Closest Approach". Figure 8.3 shows a front view of the active head designed and implemented in our laboratory where $\mathbf{H}$ = the vertical distance of the head center above the ground plane, $\mathbf{I}$ = the horizontal distance between the left and the right vergence axes, and c = the offset along either vergence axis between the intersections with the elevation axis and the camera optic axis.

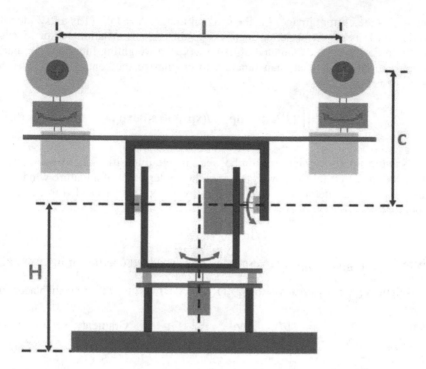

Fig. 8.3. Front view of the active head designed in our laboratory with sensor-scanning-pan-tilt system, two webcams and four servo motors for individual control of four degrees of freedom (pan control, tilt control, left vergence control and right vergence control)

Once new landmarks or image patches are identified and tracked between left and right camera images they can be initialized in the map utilizing the usual procedure of new landmark initialization in our EKF-based SLAM algorithm. Similarly, identification and tracking of image patch(es) in left and right camera images, which was(were) also previously identified in images acquired for a past position of the robot, will constitute the re-observation step of our EKF-based SLAM algorithm. In this step, where the estimated position of this landmark is calculated according to the usual "Predict" step of the Kalman filter, it is further refined by performing the corresponding "Observe and Update" step of the Kalman filter algorithm.

The steps followed for this vision-sensing based real-life implementation of EKF-SLAM algorithm is shown in Algo. 8.1. Here it can be seen that the robot is asked to move through some waypoints and it is directed to build a map of its surrounding. To perform this function, the robot is moved by a specified distance and it grabs several image frames to perform landmark observation as well as its own localization simultaneously. To build a map for both environment ahead of the robot, environment to its left and environment to its right, it is taking image shots both for 0° angular position of the pan-angle, for +θ° angular position of pan-angle and for -θ° angular position of pan-angle. Hence during the "observe" step of the EKF the robot identifies and acquires feature(s)/landmarks(s) from environment straight ahead of it, from environment to its left and from environment to its right. This procedure of moving the robot ahead, performing the "predict" step, using vision sensors in several pan directions to acquire and track landmarks, and to perform "correct and update" step of EKF algorithm is performed in an iterative fashion, until the last waypoint is reached. The map built in the last iteration is utilized as the final map built by the robot, to be used for some future tasks in the same environment.

Step 1. Specify the waypoints through which the robot should navigate and initialize the robot pose.

Step 2. Move the robot by a specified amount and perform the "predict" step of EKF.

Step 3. Grab image frames from continuously running video sequences in left and right camera, for $0°$ angular position of the pan-angle, and perform feature identification, tracking and distance calculation of the tracked feature(s) from the robot.

Step 4. Repeat Step 3 for $+\theta°$ angular position of pan-angle.

Step 5. Repeat Step 3 for $-\theta°$ angular position of pan-angle.

Step 6. For new feature(s)/landmark(s) observed in step 3 - step 5, initialize them in the map.

Step 7. For those feature(s)/landmark(s) observed in step 3 - step 5, which were observed earlier, perform the usual "observe and update" step of EKF, to refine the map already built.

Step 8. Perform step 2 – step 7 until the robot reaches the last waypoint specified.

Step 9. Store the last map built by the robot as the final map built for the environment.

Algo. 8.1. The Real EKF-based SLAM algorithm implemented for the KOALA robot, using vision sensors, in an indoor environment (in our laboratory)

8.4 Real-Life Performance Evaluation

As we have mentioned previously, the KOALA robot is a 32 cm x 32 cm, six wheeled, and differential drive vehicle manufactured by K-team, Switzerland. It has already been mentioned that in KOALA, the hardware control is performed by an on- board microprocessor (16MHz Motorola 68331@ 22MHz) [8]. To add the four degrees of freedom to the robot system for pan, tilt, left vergence and right vergence control, we have developed a PIC 16F876A micro-controller based system that, in interrupt-driven mode, works in conjunction with the Motorola processor of the KOALA robot, in master-slave configuration. The development of such a PIC micro-controller based system for interfacing external add-on peripherals with a real mobile robot, is really helpful for adding flexibility for real life applications and this development was discussed in detail in chapter 2.

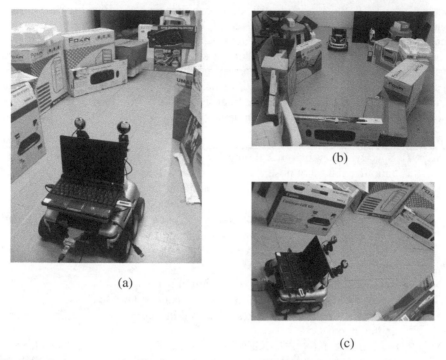

(b)

(a)

(c)

Fig. 8.4. The environment created through which the robot navigates and performs EKF-SLAM algorithm

Figure 8.4 shows the indoor environment created through which the robot is asked to navigate through several specified waypoints and build a map performing vision-based SLAM algorithm. To judge the performance of the system, a grid containing 100 squares was drawn on the maze with each square having a dimension of 20 cm × 20 cm i.e. a navigation domain of dimension 2 m × 2 m was explored.

Figure 8.5 shows the GUI-based software developed in our laboratory for real-life execution of the EKF-SLAM algorithm. Different frames in Fig. 8.5 show the landmarks identified during several iterations for incremental map building employing the EKF-SLAM algorithm and incorporation of these landmarks in the stored map. The "green line" shows the ideal path joining the waypoints through which the robot is asked to navigate. The "light blue triangle" represents the initial, starting pose of the robot and, as can be seen in Fig. 8.4, this initial pose for our implementation is considered as: $(z, x, \phi)^T = (-100, 0, 0°)^T$. For this real-life implementation here, the notations z, x and ϕ are chosen in conformation with the notations used in [3] and hence the z-direction and x-direction correspond to the x-direction and y-direction respectively, as specified in our theories before.

During its navigation, the robot identifies landmarks in its surrounding environment and initializes their positions or refines their positions in the map. As the robot keeps moving forward, the number of landmarks identified, and hence, the size of the map, increases. The "red crosses" in the map show the 2D positions of the landmarks identified. Figure 8.5(d) shows the final map constructed at the end of the test-run of the KOALA robot.

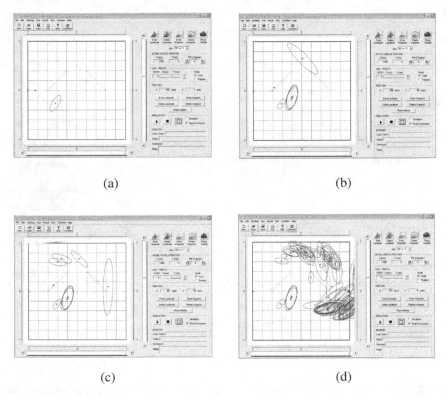

(a)   (b)

(c)   (d)

Fig. 8.5. Real-life landmark identification for map building in different steps of EKF-SLAM algorithm

Figure 8.6 shows the GUI-based form developed for capturing image frames in real-life, for some representative positions of the KOALA robot and demonstrating the performance of feature extraction and tracking algorithm, for meaningful identification of landmarks. The image patches identified in "red squares" are identified as new potential landmarks and the image patches identified in "green squares" are identified as re-observed landmarks. The form also displays the 3D distance calculated for each landmark tracked, from the robot.

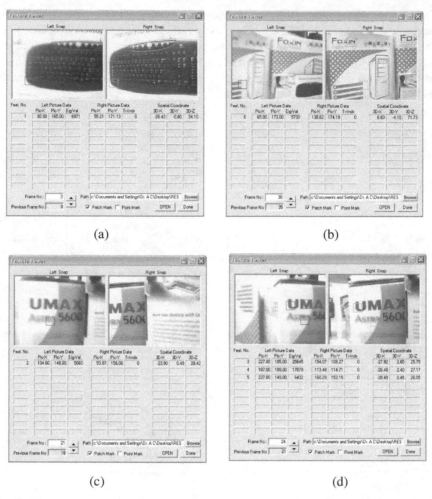

(a)   (b)

(c)   (d)

Fig. 8.6. Sample examples of results of feature extraction, feature tracking and 3D distance calculation of the tracked features from the robot, for some representative positions of the KOALA robot, during its test run in the environment

Figure 8.7 shows three sample situations of identifying and tracking features/landmarks in real environments. The "green line" on the maze and in vertical direction and the "red dots" help in pointing the actual landmark in the environment and in obtaining its true position. The hollow circle drawn in "light blue" shows the actual object corresponding to an image patch identified in the environment. The estimated positions of these landmarks in the map built, shown earlier in Fig. 8.5, show that there are small discrepancies between the true 2D positions and the estimated 2D positions for most of the landmarks in the map. However this is always understandable and can be appreciated for real-life experimentations. Table 8.1 shows these true and estimated positions, for the three sample landmarks under consideration, as shown in Fig. 8.7.

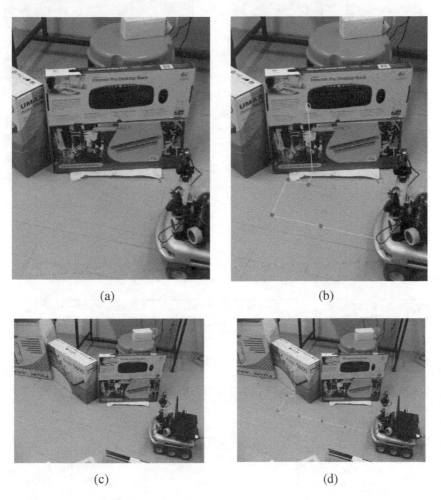

(a)   (b)

(c)   (d)

Fig. 8.7. Three sample situations of identifying and tracking landmarks in real environments

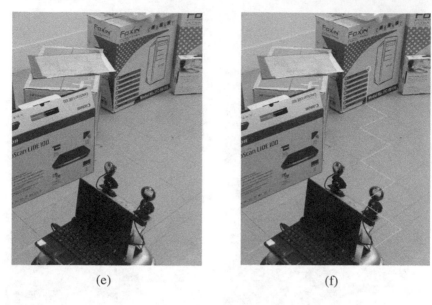

(e)   (f)

Fig. 8.7. (*continued*)

Table 8.1. Performance comparison of the EKF-SLAM algorithm employing vision sensing, for three sample real-life landmarks, as shown in Fig. 8.7

Sl. No.	Landmark Description	Estimated Position (cm)		True Position (cm)	
		z-coordinate	x- coordinate	z-coordinate	x- coordinate
1.	Landmark in Fig. 8.6(a) and Fig. 8.6(b) (bottom left corner of the keyboard image)	-43	-26	-47	-27
2.	Landmark in Fig. 8.6(c) and Fig 8.6(d) (corner of the letter 'A' in UMAX box)	-18	-18	-10	-23
3.	Landmark in Fig. 8.6(e) and Fig. 8.6(f) (top right corner of the thick red line in the FOXIN box)	4	70	2	74

8.5 Summary

In this chapter we described the theories of and successfully demonstrated a real-life implementation of the simultaneous localization and mapping problem (SLAM) of mobile robots for indoor environments, utilizing two web-cam based stereo-vision sensing mechanism. The system showed a successful implementation of an algorithm for image feature identification in frames grabbed from continuously running videos on two cameras, installed on the active head integrated in-house with KOALA mobile robot, tracking of features/landmarks identified in a frame in subsequent frames and incorporation of these landmarks in the map created, utilizing a 3D distance calculation module implemented in real-life for calculation of co-ordinates of landmarks in WCS on the basis of the distances calculated of the landmarks from the robot frames. The system could be successfully test-run in laboratory environments where our experimentations showed that there are very small deviations of the estimated landmark positions determined in the map from the actual positions of these landmarks in real-life. It is hoped that such successful implementations will inspire many readers to implement similar meaningful map building systems for more complex environments and also in outdoor situations.

Acknowledgement. The work reported in this chapter was supported by All India Council for Technical Education under RPS scheme (Grant No. 8023/BOR/RPS-89/2006-07).

References

[1] Dissanayake, M.W.M.G., Newman, P., Clark, S., Durrant-Whyte, H.F.: A solution to the simultaneous localization and map building (SLAM) problem. IEEE Tran. Robotics and Automation 17(3), 229–241 (2001)

[2] Smith, R., Cheeseman, P.: On the representation and estimation of spatial uncertainty. International Journal of Robotics Research 5(4) (1986)

[3] Davison, A.J.: Mobile Robot Navigation Using Active Vision. PhD Thesis, Univ. of Oxford (1998)

[4] Bailey, T.: Mobile Robot Localization and Mapping in Extensive Outdoor Environments. PhD Thesis, Univ. of Sydney (2002)

[5] Davison, A.J., Murray, D.W.: Simultaneous localization and map-building using active vision. IEEE Tran. Pattern Analysis and Machine Intelligence 24(7), 865–880 (2002)

[6] Chatterjee, A., Matsuno, F.: A neuro-fuzzy assisted extended Kalman filter-based approach for Simultaneous Localization and Mapping (SLAM) problems. IEEE Trans. on Fuzzy Systems 15(5), 984–997 (2007)

[7] Chatterjee, A.: Differential evolution tuned fuzzy supervisor adapted extended Kalman filtering for SLAM problems in mobile robots. Robotica 27(3), 411–423 (2009)

[8] KOALA User Manual, Version 2.0 (silver edition), K-team S.A., Switzerland (2001)

[9] Nishimoto, T., Yamaguchi, J.: Three dimensional measurements using fisheye stereo vision. In: SICE Annual Conference, Japan, pp. 2008–2012 (September 2007)

[10] Brichfield, S.: KLT, An implementation of the Kanade-Lucas-Tomasi feature tracker, http://www.ces.clemson.edu/~stb/klt

[11] Shi, J., Tomasi, C.: Good Features to Track. In: IEEE Conference on Computer Vision and Pattern Recognition (CVPR 1994) Seattle, pp. 593–600 (June 1994)

[12] Marr, D., Poggio, T., Ullman, S.: Bandpass channels, zero-crossing, and early visual information processing. Journal of the Optical society of America 69, 914–916 (1979)

[13] Moravec, H.: Obstacle avoidance and navigation in the real world by a seeing robot rover. PhD thesis Stanford University (September 1980)

[14]   Yamaguti, N., Oe, S., Terada, K.: A Method of distance measurement by using monocular camera. In: SICE Annual Conference, Japan, pp. 1255–1260 (July 1997)

[15]   Chatterjee, A., Ray, O., Chatterjee, A., Rakshit, A.: Development of a Real-Life EKF based SLAM System for Mobile Robots employing Vision Sensing. Expert Systems with Applications 38(7), 8266–8274 (2011)

[16]   Chatterjee, A., Singh, N.N., Ray, O., Chatterjee, A., Rakshit, A.: A two-camera based vision system for image feature identification, feature tracking and distance measurement by a mobile robot. International Journal of Intelligent Defence Support Systems 4(4), 351–367 (2011)

Index

CPSIA information can be obtained
at www.ICGtesting.com
Printed in the USA
LVHW06s1032010718
582387LV00005B/28/P

9 783642 426704